BELIZE CHRONICLES

1991–1995

TERRY VULCANO

Black Rose Writing | Texas

ISBN: 978-1-68433-329-5
PUBLISHED BY BLACK ROSE WRITING
www.blackrosewriting.com

Printed in the United States of America
Suggested Retail Price (SRP) $18.95

Belize Chronicles is printed in Chaparral Pro

TO MY FRIENDS FROM TWENTY-FIVE YEARS AGO
WHO IT WAS JOY TO RECONNECT WITH
AND SHARE THESE STORIES AGAIN.

Significant City States of the Maya

- Mega Maya sites (100,000+ residents)
- Major Maya sites
- Minor Maya sites

Santa Rita
Corozal
Cerros

Noh Mul

Cuello

Calakmul

MEXICO

La Milpa

Altun Ha

Lamanai

El Mirador

Chan Chich

Belize City

Uaxactun

El Pilar
Buena Vista
San Ignacio
Cahal Pech
Xunantunich
Pacbintun
Las Ruinas

Tikal

Naranjo

Yaxha

Flores

Caracol

Dangriga

BELIZE

Num Li Punit
Lubaantun
Uxbenka
Punta Gorda

GUATEMALA

Pusilha

Map layout by D. Driediger

BELIZE CHRONICLES

1991–1995

PREFACE

When I went to Belize and sent stories to friends, I never thought of writing a book. After receiving positive feedback urging me to consider publishing the events of daily life, I retained duplicates of my correspondences. In retelling the events, I have attempted to bridge gaps with postscripts.

Previously (1982-1985) I had been a volunteer in the northern Kalahari area of Botswana helping single entrepreneurs. In Belize, it was helping groups (cooperatives).

The subject matter is diverse, covering things like travels, events on the farms, thoughts I had and environmental issues. This anthology of stories is intended to give a sense of the experiences of living and coping in Central America; what the country and surrounding area was like; what kind of things happened and how things were; how I felt, thought and responded to what I was going through.

Terry Vulcano
November 2018

BELIZE BACKROUND

Introduction

Belize, on the east coast of Central America, is also considered a Caribbean country. It is grand for vacations due to its Maya ruins, rainforest excursions and coral reef. At 180 miles long and almost 70 miles wide it has an area of about 8,800 square miles. Its population is around 400,000, double what it was in 1991. It has the lowest population density in Central America. Mexico is to the north with Guatemala to the west and south. Honduras is a ferry ride away to the south

The north of Belize is mostly swampy coastal plains, with heavily forested areas. The south contains the Maya Mountains. The highest elevation in Belize is 3,688 feet. Three fifths of Belize is covered by forest, with one fifth of the land in agriculture. Some of the forests are in protected areas. A quarter of forests outside of protected areas have vanished since 1980.

Guatemala considers Belize as part of its territory, whereas Belize deems itself an independent nation with Queen Elizabeth II as its head of state. While English is the official language Belizean Creole is widely spoken with Spanish the second most common language. Belize is divided into six districts: Corozal in the North, bordering Mexico; Orange Walk to the Northwest, bordering Mexico & Guatemala: Belize (City area) on the north coast; Stann Creek below that; with Toledo District in the South, bordering on Guatemala; and Cayo to the west, also bordering on Guatemala. It was in Cayo that I was involved with economic development projects.

Climate

Belize has a tropical climate with wet and dry seasons. Temperatures hardly vary (average 75 °F in winter to high 80s °F in summer). I noted this constant heat, day or night, my first year in Belize, having been used to extremes in Canada. There was only one day, out of four years, Christmas morning 1994, that I wore a long sleeve shirt. Even in the Kalahari temperatures went below freezing in June. Temperatures are higher inland, with mountain areas lower, more comfortable. Rainfall varies from four feet annually in the north and west, to fifteen feet in the south. This is befitting rainforest growth. However there can be droughts in the region like in 2019 and times past.

Flora & Fauna

Belizean rainforests are home to jaguars, tapirs, coatamundis, howler monkeys, spider monkeys, armadillos, peccaries, other mammals, a variety of snakes & tortoises, 500 species of fish, 600 species of birds and 5000 species of plants. The national flower is the black orchid. Mahogany is the national tree. The national animal is the Tapir and the national bird is the Toucan.

The Economy of Belize

Sugar constitutes half of Belize's exports, while the banana industry employees the most. There are many citrus orchards throughout the country, some adjacent to orange juice making facilities. Soya bean farming has seen growth in the last quarter century. There is still some logging. Tourists make a large contribution to the economy.

Tourism

Belize is no longer the Undiscovered Country. It attracts a quarter of a million visitors a year. In thirty years it went from the tenth most important industry to first. This small country became a desirable destination on account of its steady warm climate, barrier reef, islands called Cayes (pronounced keys), fishing, scuba diving, snorkelling, kayaking, rafting, rainforest hiking, wildlife reserves, bird watching, caving and has a dozen Maya sites open to visitation. It helps that English is the national language. Belize hosted the First World Congress on Tourism and the Environment in 1992.

The Belize Barrier Reef is a UNESCO-recognized World Heritage Site. It is part of the Mesoamerican Barrier Reef System that runs from Cancun to Honduras. It is the second-largest in the world, behind Australia's Great Barrier Reef. The reef is threatened by tourism, fishing, oceanic pollution and global warming (higher ocean temperatures causes coral to die and turn white). Two-fifths of its coral reef has been marred. Belize was the first nation on the planet to ban bottom trawling. It also put a moratorium on offshore oil drilling near the barrier reef.

Two major annual sporting events have gained international interest. At Easter there is The Cross Country Cycling Classic (from Belize City to San Ignacio & back). Also in the spring there is the four day La Ruta Maya Belize River Challenge, (canoeing from San Ignacio to Belize City).

See **Appendices** for driving mileages from Belize City to Caracol and Tikal. Community names are interesting and position along the highways may help orient readers to some of the stories. The chart also allows for taking a vicarious tour along Belize's western highway with directions to Tikal.

1991 – THE PRE-FORMATIVE PERIOD

I NEVER WANTED
TO LIVE IN THE JUNGLE

INITIAL TOUR

My host organization was the Cayo Centre for Development and Cooperation (CCDC). They were headquartered in San Ignacio Belize and helped with projects throughout Cayo District with its western edge bordering on Guatemala. The Coordinator spent a day taking me around to a few of the groups CCDC worked within the district. We used the white Toyota 4WD, which only had a few dents at that time. We headed south along the Cristo Rey Road, a bumpy, twisty gravel road with which I became very familiar.

Five miles past the village of *Cristo Rey* (Christ King) and eight miles beyond San Ignacio (a 45 minutes drive) was the town of San Antonio. Situated here was a group of Maya farmers with a communal pig-rearing project.

The project was financed by Canada Fund and included a bio-gas plant. Executive members of the group demonstrated how their fridge, stove, and lighting worked using the gas by-product. I learned more about pigs than I ever cared to know. San Antonio also boasted the Garcia Sisters Museum. They were not a CCDC project, but they were a unique enterprise. The five Sisters had established, and ran, the renowned Garcia Sisters Museum and craft shop. In the, recently opened, exhibit area, were exquisite slate carvings they had created in their father's backyard.

He was sceptical, at first, of his daughters spending time on the carvings until he saw their value; he built them a beautiful circular rondavel like structure where the Sisters displayed their top quality craftwork for sale.

Past San Antonio is the old Maya site of Pacbintun. The caretaker gave us a "three dollar" tour. He pointed out the ancient causeway that extends past the gravel road. The site, known for its cache of ancient instruments, had been excavated by Dr. Pendergaast, a Toronto based archaeologist.

About one mile further we reached a T-junction. The right led to Mountain Pine Ridge. We turned left. The bumpy trip continued through scenic country. We met members of the Las Vegas group at the next stop. They also wanted to raise pigs and were waiting funding. After heading north, we reconnected with the Western Highway at Georgeville.

We headed towards San Ignacio (also known as Cayo) but turned north at the British Army camp to take the ferry across the Belize River. It was a hand-cranked ferry that could hold three cars at a time. The crossing took 20 minutes. Anyone was welcome to take a turn at the crank if they were not happy with the speed of the ferry.

This area was flatter and mostly cleared land in contrast to the hills and jungle-like vegetation around Mountain Pine Ridge. The road wound through maturing fields of corn.

We saw the location for the proposed facility for The Progressive Group of Santa Familia (SFPG). They grew maize and soya beans. They had their own tractor for planting and harvesting - hard work without the proper machinery. Cahal Pech hill could be seen about two miles away but was seven miles by road. Where the Mopan and Macal Rivers meet to form the Belize River was the community of Branch Mouth,

The Branch Mouth Women's Group (BMWG), a group of seven women, raised poultry on a rotational basis. It takes seven weeks for chicks to turn into broiler size hens. They raised chickens so that each week a different member's chickens matured for sale at the market. Working as a group, they slaughtered, plucked and packaged the chicken. They did not make significant money for all their efforts but they liked having their own project, and the community liked having the project in the village. I was to learn plenty about chickens.

At Branch Mouth is the Las Casitas Resort. We could spot iguanas sitting in the trees across the river. Three miles further is the community of Bullet Tree Falls where a bridge crosses the Mopan River. In the community of 1,000, CCDC was helping the Crystal Hatcher's Association (CHA) start an incubator-hatchery operation. A facility had been built; they were waiting for further funding for equipment. It was three miles more back to San Ignacio, and that made for a full day.

ON THE BUS

December 23, 1991

Yesterday I journeyed by bus from Cayo (San Ignacio). It was a four-hour trip - one hour waiting for a bus that leaves every fifteen minutes (eight went the other way in the same time period). At first, I had a good seat at the front. With the bus stopping every two to three blocks (only for the first thirty miles) I was bumped and banged every time someone boarded.

When a mother with her baby alighted, I gave up my cherished seat. Now standing, I was awarded with an eyeball view of the blaring roof speaker and an occasional glimpse of the road as it passed underneath. As others piled on, I was unceremoniously shoved towards the back - separated from my travelling bag that was in the front rack right by the door. It was easy grabbing for some departing passenger.

Six miles down the road a seat became available. I was thankful to sit again and at least see out the side window even if I did not feel the breeze from the open door. Another mile, another mother with another baby, resulting in another gesture of giving up my seat. I offered to hold the baby - but the baby would have nothing to do with that!

By the time we arrived at Belmopan, there were about ninety of us on the forty-four-passenger bus. At the capital, one-third of the way to Belize City, one - third of the ninety got off. I located another seat halfway towards the back. About fifty-five people got on making it 115. I was still separated from my bag.

There was a young boy, seven years old on my right; a girl of about eight slept against my left shoulder; a boy of three sat in my lap, and a baby of ten months was in my arms. I could not give up my seat if I wanted to. I was further from my bag than ever.

At first, all was calm. The bus made good time with a few stops, a few potholes - too calm! I worried "What if a child woke up and wanted their mother?" What if some mother does not claim her kid at trip's end? I did not know who belonged to whom!

The three-year-old stood up and began to cry. A pair of arms reached out, and he was secure again. About halfway to Belize City, the boy on the right got off, on his own. I crammed against the window; two more kids filled up the space. The girl and baby still slept. I wondered what was going to happen when the Creole baby woke up to find a bearded white man holding her. Everything turned out fine. No kids went unclaimed, and my bag did not go missing.

History

Spanish explorers proclaimed Belize as part of the Spanish empire. However they did not settle the area due to indigenous hostility and apparent lack of potential in the region. English pirates colonized the coastal area, seeking places to cut mahogany trees, with a permanent English settlement in place by the early 1700s.

When war broke out with Britain, the Spanish attacked the area. The Battle of St. George's Caye was the last confrontation. The Spaniards were warded off on September 10, 1798, which became a national holiday in Belize.

British Honduras

Slavery was abolished in the British Empire in 1833. Being restricted from buying land in the settlement, the former slaves continued to cut timber for low wages. In 1862, Great Britain declared the area a British Crown Colony, named British Honduras.

The colony relied on the mahogany trade from the 1860s until the 1900s. The demand for timber fell during The Great Depression. Economic conditions did not improve until World War II.

The **Caste War of the Yucatán** (1847–1901), a revolt of the Yucatan Maya, against the European-descended population, involved the Mexican War of Independence, The Mexican-American War and the Mexican Revolution. The fifty-four year conflict enmeshed Belize in a number of ways. There were battles and treaties among the Maya, Mexicans, Guatemalans and British. Yucateros, who had economic control of the region, fought the independent Maya. Losses have been estimated between 50,000 and 300,000 dead. The war had official endings in 1855, 1901 and 1915 but skirmishes continued until 1933.

New Spain, including the Yucatan, had an official sixteen level caste system. As a simplification there were: those born in Spain at the top; followed by those of Spanish descent; then *Mestizos*, having a mixture of Spanish and native blood; followed by *Hidalgos*, natives who helped the Spanish; and lastly *Indios* who were the majority of the population. The elites treated the Maya poorly and underpaid them. This caste system did not exist in British Honduras but there were slaves.

Following the Mexican War of Independence (1810-1821) the Yucatan declared independence and subsequently joined the Mexican Empire. However the frontier areas resented Mexico City's centralized decision-making. Several provinces revolted, including Guatemala in the south and Texas in the north. To shoulder the costs of the internal conflict the Mexican government imposed taxes, leading to further revolt.

Revolt appealed to the Maya. They were told their land would no longer be exploited if they joined. With their support the Yucatan declared itself independent of Mexico in 1841. Although the Mexican government blockaded the Yucatán, their forces were thwarted in attempts to take the cities. The Mexican-American War (1846-1848) made the belligerents afraid of an American takeover so fighting abated.

Then a clash arose against the taking of more land. The Yucatec's expanded sugar and agave plantations encroached on Maya communal land. Maya towns were ransacked and people indiscriminately killed. The Maya responded to the massacres by killing non-Maya. Both Mestizos and Maya fled to British Honduras. By 1848, the Maya forces had taken over most of the Yucatán. The Maya lost the advantage when they left to plant crops.

Due to Mexican troops helping the Yucateros, the Maya were limited to the southeast of the Yucatan. In the 1850s the United Kingdom gave recognition to the Maya state (because of trade with British Honduras). However increased investment in Mexico led to an 1894 treaty, recognizing Mexico's jurisdiction of the Yucatán. This formalized the northern border for British Honduras.

The insurgent Maya battled the Mexicans, the Cruzob (Maya devoted to the Catholic Church) and forces of British Honduras. They defeated British troops on December 21, 1866, at the Battle of San Pedro. This is not a national holiday in Belize. The Maya briefly took Corozal Town in 1870. Then lost their leader at the Battle of Orange Walk. Remains of two forts from that time can be viewed in downtown Orange Walk. Because of their leader's death, the Maya made peace with the British, only to have a timber and cattle ranching company forcibly remove them, years later. Some of that land became a Nature Conservancy in the 1990s. The Mexicans reneged on an 1884 treaty signed in Belize City and hostilities resumed.

New economic factors and social changes coming out of the 1910 Mexican Revolution reduced enmities. Reforms helped remove the issues that had been the cause of the wars.

Hurricanes

Because Belize lies on the Caribbean coast and that area is low lying Hurricanes have had devastating roles. Although not as frequent as the ones in the Gulf Coast or along the Florida coasts they have had great impacts in Belizean history. Noted below are eight of them since the 1930s:

Year	Name	Damage
1931	noname	destroyed two-thirds of the buildings in Belize City, killing more than 1,000 people.
1955	Janet	levelled the northern town of Corozal
1961	Hattie	struck with 185mph winds bringing 12' storm tides devastating Belize City this led to establishing the capital inland
1978	Greta	caused more millions of dollars in damage along the southern coast
2001	Iris	a 145 mph Category Four storm, demolished homes and crops in Monkey River Town, with destruction as far as Punta Gorda
2007	Dean	a Category 5 storm, landed north of the border, causing extensive damage in northern Belize
2010	Richard	made landfall 20 miles south of Belize City affecting housing and crops
2016	Earl	flooding in Belize City

It may be noted that prior to the Year 2000 there was only about one every two decades. The frequency has increased to two per decade, since the start of the new millennium. One never knows when a hurricane will hit and thus my story "Waiting for the Hurricane."

1992 – The Formative Period

Tequila Express to Tikal

NOTES ON WORK

January 1992

After a couple of months, I have a better grasp of work and the tasks at hand. The Cayo Centre for Development and Cooperation (CCDC) works with eight different groups:

Work is going well. I enjoy the analyzing, evaluating, recommending, organizing and implementing. We are taking a fresh approach and meeting every Friday afternoon to plan for the next week. The CCDC Board has given staff permission to share coordination responsibilities. Board members were supportive of our ideas.

The CCDC affiliated groups in 1991 were:

21 de Septiembre -a group located near the Guatemala with timber and agriculture projects;

San Ignacio Women's Association -SIWA with sewing, baking and daycare planned projects for San Ignacio (Cayo)

Crystal Hatcher's Association -CHA with a facility for a chick hatching project in Bullet Tree Falls west of Cayo;

Branch Mouth Women's Group -BMWG with a chicken raising project three miles from Bullet Tree Falls, one mile from Cayo but distanced by the Mopan and Macal Rivers;

Progressive Group - located in Santa Familia, another mile past Branch Mouth with projects in crop raising focused around their tractor equipment;

Lost Tambos Group - is ten miles past the Mennonite community of Spanish Lookout. Projects include a school, a road, pig-raising, and chicken raising;

Las Vegas Group -formed to raise pigs, they are located about four miles from the village of Georgeville on the road to the Maya Mountains;

San Antonio Peanut and Pig Rearing Group -located in the Maya village of San

Antonio, eight miles south of San Ignacio. Projects included pig rearing and peanut production.

Postscript 2019*: On a visit back to the only project I could locate was the Western Leather Cooperative with one of its original members. I was told the San Ignacio Women's Association still had day care, but I could not locate the facility despite having helped build it.*

CAVES AND FALLS

January 12, 1992

The Mountain Pine Ridge area is thirty miles south of San Ignacio. Access was indeterminate due to impassable roads in the rainy season. The area contains spectacular caves, tremendous waterfalls and swimming spots.

Arrangements were made to use the CCDC truck. One of the staff wanted to take her family and friends to see Mountain Pine Ridge, so we shared the cost. Getting to Cristo Rey, three miles south of town, was slow going - the road was bad. It became worse as we travelled to San Antonio, five miles past Cristo Rey. The climb to the Mountain Pine Ridge was rugged, but the road improved once we passed the Mai Gate where a guard determines nationality, destination, and intentions. Our goal was the forestry camp of Augustine, near the caves and pools.

The first stop was the Hidden Valley Falls. The falls are known as One Thousand-Foot Falls but are actually closer to 1,600 feet high. We had to wait for the mist to clear for a glimpse of the falls. The falls were spectacular as they dropped to an invisible stream hidden in the rainforest below.

The caves were incredible. We spent eighty minutes crawling around a series of tunnels, caverns, chambers, and passageways. We did not find the end - we returned for fear of getting lost with only a penlight flashlight between the two of us.

Rio-Frio Cave was fantastic! Two openings with the big entrance, 100 feet wide by 70 feet high, led into a tunnel about 1500 feet long with a 90-degree gradual turn. If not for the trees at the entrance, a plane could fly through! It was sensational inside. Stalactites hung from the immense ceiling. There were ledges on each side with the Rio Frio running between them.

It took some acrobatics to navigate to the far side, to inspect the large Icing Cake, as I called it. I explored a dark ledge and fell into a hole. The drop would have been thirty feet to the running water below. One leg and most of me dangled while my knee and one elbow were wedged in the opening. Multiple bruises and abrasions were noticeable after my escape; I was thankful for not falling the whole way. Usually, people are led by a guide who warns visitors about the hole.

After the caves, we went to the Rio-On Pools, a series of pools that empty into one another. This was a place for swimming, soaking, and tanning. Some of the nine or so pools are connected by rocky slide ways the more adventurous use as a waterslide. Waterfalls, about ten to fourteen feet high connect other pools.

At one point I noted a girl caught in the flow and she was slowly but surely slipping toward one of the larger cascading waterfalls. I dove in and brought her back to safety but then was caught in the grip of the flow heading over the falls. I looked towards the girl and she gave me a glance to say, "I never asked you to help me."

I orientated myself to go over feet first. At the last possible moment, I was able to extract myself from the perilous situation by standing up on the rocks at the top of the falls. Two near-escapes in one day.

A TEARFUL ENCOUNTER

January 23, 1992

The rains are back. I started to work carrying an umbrella in the heavy drizzle. A little Mestizo girl was coming the other way. Tears streamed down her face in addition to the rain washing her hair and soaking her clothes. She was oblivious to the rain but distressed about something. I crouched down and offered her the covering of the umbrella. "What's the matter?" I asked.

She just looked at me with her big round eyes, holding back a sob. Maybe she did not understand English. I tried Spanglish (a butchering of English and Spanish). "Are you going to school?" She shook her head no. "Are you going home?"

"Mommy," she sobbed.

"Do you want me to go with you?" I offered.

Again she shook her head. I dabbed tears from her eyes, and she went purposefully onward. I never learned what was troubling her.

21 DE SEPTIEMBRE GRUPO

January 23, 1992

Today I visited the 21 de Septiembre Grupo (named for Belize Independence Day) at their project site, near the Guatemala border. They were raising crops in an inaccessible area. Access was by 4WD in good weather. I doubted we would make it in the rain. The Field Officer, for the group, thought there would be no problem, as it had been dry this week.

There was a road being constructed to facilitate the new power dam. It was under heavy construction, and the combination of dirt, boulders, and steep inclines made four-wheel drives a necessity to get through (either that or travel by horse). We picked up two group members, in the border town of Benque Viejo. We also took supplies to deliver to their agriculture camp. Travel was slow - 15 mph. The baldness of the tires prevented getting up the path to the site. We walked the last three miles carrying the supplies. It was still raining.

The rain forest loomed on either side of the trail. About two miles in we came to a clear-cut area devastated by an American who had been granted logging rights. The 21 de Septiembre Group only has a permit to do selective logging where the road will go through. On the way in we met others on the way out.

Apparently, there is a dispute over who is taking the easy-to-get timber. A reconciliation was made. In this area the Cooperative members planted corn and peppers - only harvested by hand because of the hilly terrain.

Numerous birds flitted from tree to tree and butterflies fluttered about. The group's camp is a series of unfinished thatched shelters. Lunchtime arrived.

BEANS IN MY STOMACH! BEANS ON MY SHOULDER!

I hate baked beans, but the hiking gave me an appetite. I gladly accepted the plate of beans and rice prepared by Don Fabio, their kindly cook, nicely flavoured by their local pepper sauce. It is like camping out; they do it week after week. After eating we discussed the cooperative.

One item was on the purpose of a cooperative. In their December evaluation, they had mentioned the advantages of being their own boss. But now they were employing others to do the hard work - carrying out the cut timber. If there was enough work to employ others, perhaps they should consider allowing them to join the cooperative.

Most of the members were Guatemalans; some still lived in Guatemala in the border town of Melchor. Members agreed with this but had difficulties collecting membership dues. This was a matter for future resolution - how to come up with a system that would be fair to the group who had done so much work preparing the land.

Another issue was reforestation. Members wanted to start replanting mahogany to replace the trees cut down. I would research information for such a program. Another area of interest was organic fertilizer. The field officer promised to investigate possibilities. After the discussions, I took a walk around to find that one of their thatched structures is set atop a large Maya temple.

If I was impressed with this, they assured me, there was another one twice as large a few miles further south. When I had checked the archaeological listings of known sites neither seemed to be mentioned.

On the trip out, we each carried 100-pound sacks of beans on our shoulders, gathered from their harvest, for sale in the Benque shops. The heavy drizzle continued. There was no place to put the sack down to adjust my boots. The boots were ruined.

The road was much mushier for the return trip. One long incline was too steep for the truck to negotiate. Five passengers got out to push. Progress was slow but steady. The rain has become a mild downpour. A short distance from the top we rested. Instead of driving I offered to trade places with one of the ones pushing.

In the brief moment, we switched the vehicle slipped back hundreds of yards; the wheels locked with brake pressure but the running water made the surface slippery as ice. We regained lost ground but not our former position.

We blocked the wheels with large rocks. We threw loose rock chips, about thumb size, on the surface ahead. The wheels grabbed greedily on the new traction, and we made it over the hurdle. There was mud on the windshield and mud on my glasses, but we were on our way out.

TEQUILA EXPRESS TO TIKAL

February 22, 1992

Phil Watson visited from Dundas, Ontario. We had been in Botswana together, from 1982 to 1985. Phil had set up a shoe factory in Kanye, just south of the Kalahari Desert proper, while I was stationed in Ghanzi District, in the northern Kalahari. While in Ghanzi, Phil had taught leather working courses for entrepreneurs. He was going to put on a brief course for the leather group in Belize. Before the course we planned a trip to Tikal, the mega Maya site in Guatemala. It is fifty miles to the west, as the quetzal bird flies. We rose at 4:00 a.m. to catch the 5:00 a.m. bus at the Guatemala border. We had to cross when the border was closed.

We acquired visas the day before, as the border crossing did not open until 7:00 a.m. We walked across the bridge in the dark. Crossing in the opposite direction were smugglers, out of Guatemala, with bag loads of goods to sell in Belize. Once over the bridge, I noticed Guatemalan soldiers observing us with the barrel of their machine guns pointed our way. I waved; they did not shoot.

The bus was punctual; it left at 5:05 AM. We were Peten bound. There were eight other passengers on the bus. It reeked of tequila and gin. I kept sneezing!

There was a full moon, but it offered little illumination through the thick fog. The bus trundled along at 8 mph, bump after bump after "dang" bump. Sometimes it would slow to 4 mph. Other times it got up to 14 mph. It took over four hours to go 32 miles to the spot where we could catch a taxi bus to reach Tikal turn.

I thought going 20 mph on the Kalahari rut trails was slow going. For the first hour and a half, we did not stop once - it was an express; the tequila express. Other passengers were boarding as we reached communities along the way. The bus only had seating for 32, but there were 67 of us by the time Phil and I reached the transition stop. The cost was eight Quetzalas (named for the national bird).

It had not been a boring journey. There was a military checkpoint where fifteen-year-old illiterate soldiers, with AK 47's, inspected our passports. Passengers resembled the cast of Viva Zapata.

Nevertheless, I was glad to be off and wait along a paved road to Tikal. We caught a ride on a mini-bus for 25 Quetzalas. Seventeen kilometers outside Tikal we paid 30 Quetzalas to enter Tikal National Park. Arriving at about 10:00 a.m. we tried to find a place to stay for the night. All three hotels were full. We left our bags at the entrance and went into the site.

TIKAL SITE

At the entrance were interpretative buildings holding stelae, large carved stones depicting an event from Maya history along with a huge model of the excavated portion of the site. Still, it did not give an accurate impression of the massiveness of the site. I had studied Tikal in archaeology courses and had seen slides and films about it, but none of that prepared me for how IMMENSE it was. Temples and pyramids were more incredible and fantastic than the preceding ones.

TIKAL COMPARED TO DOWNTOWN CALGARY

To furnish a sense of size and spaciousness an analogy with Calgary is used. Starting with the downtown core is the heart of the great plaza. Imagine three blocks of the eighth avenue mall and one block wide. At the east end of this plaza is Temple I. It appeared to be about the size of the Norcen Tower but located where City Hall is. At the other end is Temple II - on par with the Bay.

In the downtown area is the North Acropolis – about the volume of the Municipal Building to get a sense of perspective. Across from that, on the south side of 8th Avenue, is the equivalent to the SAIT complex but stretching the whole length of the plaza and then twice that much again. To the end of the Central Acropolis is the East Plaza - about twice the size and location of the Planetarium.

Near the location of Dean House is Temple III. It looks like the size of the twelve-story Library Tower at the University of Calgary. At the 12th Street bridge to the Zoo is the Bat Palace Complex - it is about the dimensions of the Palliser Hotel but only half as high. (There is much in between.) Temple IV looks as high as the Petro Canada Building (Temple IV is 212 feet tall but seems higher when scaling its unexcavated sides). This describes less than 2% of the site and less than half the structures in that 2%. The part of the site open to tourists continues to the equivalent of Dover in the east and Shaganappi Park west. There is an abundance of equally impressive structures going from 10th Street NW to the Zoo, two blocks either side of Memorial

Drive.

This area includes such features as "Plaza of the Seven Temples," "Lost World" and "Temple V". There is no river. In between the latter row of structures, and the first batch, are several reservoirs such as Temple Reservoir, Palace Reservoir, and Hidden Reservoir. There are tall trees and thick forest throughout so that you cannot see from one plaza to the next or from one reservoir to the next. Even when we climbed to the top of the temples, we could only discern the tops of other temples and none of the agglomeration of other structures.

There are many smaller structures and medium size complexes throughout, connected by causeways much like shopping malls. To the south are complexes F, H, & M - think of these as Chinook Shopping Centre with building groups situated in Ogden and Glenmore. To the north is a complex likened to the North Hill Shopping Centre with outlying features in Cambrian Heights and Brentwood. We did not get to the "Temple of the Inscriptions." We were bushed from the speedy encounters and limitless temple climbing. After two rainfalls and no lunch, we were ready for a break.

In the parking lot, Phil spotted a Taxi with Belize plates and procured us a ride back to Cayo for $10 US. It felt as if we had gone to heaven (riding in the back of a car for two hours versus the five it took us to reach there by bus). Tomorrow we go to Caracol (in Belize) which is supposed to be larger than Tikal. The taxi driver says we can only get there by horse. I did not believe that. *This turned out to be close to true.*

ENROUTE TO CARACOL

February 24, 1992
Caracol is a Mega Maya site in Cayo District. The road was not upgraded when we went.

Several of us went on a weekend trip to Mountain Pine Ridge. On Saturday we did the usual: Hidden Valley Falls, Rio Frio Cave, and Rio on Pools, etc. This time we had three strong flashlights for the caves and saw much more. I brought along 1600 ASA film to capture intriguing features. When we told the entrance guard we were Caracol bound; he was skeptical about us making it on 'that' road. He found it hard to believe that we had a permit, from the Archaeology Department, the first of the season issued weeks earlier.

We stayed overnight at the Augustine Lookout Station, a nice facility, situated with a commanding view atop a hill, like a mountain resort. Belize Cabinet Ministers use if for conferences, but it is available when not booked. The caretaker advised us that the archaeology crew for Caracol had gone through the day before. Their truck had to be pushed most of the way, with a Caterpillar tractor.

At the Forestry Department, we had to obtain further permission to proceed. Again doubt was expressed as to the ability of the truck to make it (just because it had four bald tires, no emergency brake, no mirrors and had to be push started should not be taken as a reflection on the vehicle's performance capability).

In the morning we were thwarted by a road crew installing a culvert. Again there was disbelief that we had a permit and the installation crew foreman gave us the once-over as if to say he could not believe the truck got this far. He told us how to detour. Unfortunately, his instructions included numerous lefts and rights at different forks in the road. What should have taken us twenty minutes took two hundred minutes. Our first foray took us to an army camp near the river - great scenery but a dead end. Our next endeavour put us at a sawmill camp not even on the map.

The camp foreman told us there was no way (without even seeing our "trusty" means of conveyance) we could make it past the Guacamallo Bridge. We were, by this point, actually heading towards Hidden Valley Falls.

Undaunted, we retraced our "spoor" until we were finally back on the main "road".

On the map, San Luis was shown as a town; in reality - an abandoned sawmill camp. On the map, Guacamallo is an abandoned camp; in reality - overgrown vegetation. The Guacamallo Bridge was the end of the road network as known to motorized vehicles.

After crossing the bridge, we had been advised variously to turn left, turn right and go straight ahead. Putting the Toyota into 4WD, we proceeded cautiously, frequently reconnoitering the path ahead. The right disappeared into the bush; straight ahead disintegrated into a pasture; to the left - a trail?

The Road is Mush; The Road is Bad

I'd hate to see this road in the rainy season. It is deeply grooved with muddy ruts with water streaming out from all sides. Calling this Chicelero trail a road was akin to elevating a cub scout to colonel. I re-engaged the 4WD and sailed through, waltzing from mud hole to mud hole, doing continual scouting to avoid the worst.

Three years of being stuck in "downpour-created" sinkholes in the Kalahari had to be worth something. We had a reprieve at a Chicelero station where they harvested the sap of the gumbo-limbo tree. They provided some of their ware - chewing gum fresh from the forest. Pat got juicy fruit, Moses double mint while Phil had spearmint. I got cinnamon ... okay, so it was just chewy.

The worst of the road was yet to come. I was amazed to have made it this far. I was about to give up reaching any further - all those doubting permit inspectors were going to be right. Again the 4WD drive pulled us through, carrying us up steep rocky inclines, over boulders, through mud-slews and past thick vegetation.

A modern, fully equipped 4WD reached us. The driver found it incredulous that we had a permit and that we had travelled this far. There were nine miles to go. He warned us there were rough spots ahead. Two hours later we arrived at Caracol.

Maureen, an interpreter from the archaeological crew, greeted us and offered to guide us.

She apologized for the unpreparedness of their group as they had just arrived at midnight. Their big army truck had broken down completely, and they were forced to walk the last two miles carrying all their provisions. We had not expected a tour; we did not mind that it would be ad hoc. They were short of water because it had been used to feed their truck's radiator. We gave them half of our thirty gallons and enjoyed the first interpretative program of the year.

It had many pyramid mounds, quantity but not quality; the majesty and craftwork were not the same as Tikal. Caracol had been into warfare; maybe that is why their structures are not as elegant.

Caana (pronounced Ka-ah-na), the biggest structure, is massive. On the top, there was a whole plaza, probably the home of the ruling elite. This was imposing.

Another impressive thing was seeing the hollow rock, discovered last year. It is sixteen inches long, ten inches wide and a foot high. Inside they discovered a spondylus shell (from Peru?). Under the shell were a big jade head carving and a pool of mercury (first time found in the Maya realm). The find dated to the Early Classic (100 BC to AD 200). Returning the way we came was slower than slow. We returned well after sunset, without incident, without becoming stuck!

Postscript 2019: *Due to robbers coming from Guatemala and holding up tourists at Caracol there have been military escorts leaving the St. Augustine camp at 9:30am on days that suspicious activity is noted.*

ANIMAL ENCOUNTERS

Possum

One morning, at the CCDC office, we discovered a baby possum in one of the large plastic garbage bins. It had probably crawled in searching for food and could not get out. The steep sides were too slippery for the poor thing trying to get out. Some wanted to kill it because they said possums eat chickens. "Well people eat chickens," I pointed out, "Do you kill them?" They told me I could keep it as a pet. Although it was very cute, I did not want to put it in captivity. It belongs in trees. I lifted out my "little buddy". It would have stayed on my shoulders. Someone went to find some string. I made a nest for it out of newspaper that it seemed to like.

Before it became too comfortable, I coaxed it out and shooed it down some timber toward the back. It slowly climbed over the timber, through the grass, around the fence toward a large tree. It was having trouble finding its way; it was just a baby. If the poor creature can survive downtown (where there are no chickens), then it should be left alone.

Parrots

Santos, a farmer at Los Tambos, caught three baby parrots to sell in town. They looked so forsaken, cramped in their tiny cage. I wanted to set them free. I thought of paying to release them, but he would just go catch some more or maybe the same ones! They were too young to be on their own.

Sharks & Tortoises

When Phil Watson was visiting in February, we went to Caye Caulker and San Pedro (on Caye Ambergris). On the wharf, at the back of a drinking establishment, was a small enclosure where six sharks and five tortoises were crammed into a confined space. Maybe they put the tortoises in with the sharks so that people would think twice about freeing them.

There was a festival going on where people were pre-occupied with painting faces or trying not to get painted. I returned at night. When no one was looking, I took an old wooden door and lowered one end into the tank for the tortoises to climb out. My plan was to lift the tortoises out one by one and release them into the sea, where no one would learn of their escape. Alas, the tortoises did not understand the escape plan and shied away from the door to freedom.

Then I had the bright idea of letting the door float in the pool thinking the tortoises would then climb on. The tortoises eyed the door from the dark recesses. I eyed the tortoises from the dark evening.

The dark sharks eyed me hoping I would get in with the door. It was a Belizean stand-off. The pina coladas had fuzzied my thinking. No further solutions presented themselves. The floating door became an unused island for the proprietor to ponder the next day.

SOFTBALL, SNAKES, POSTS AND PILAR

March, 1992

"There, intertwined on the branch ahead of me, was the thickest green vine I had ever seen -twelve inches away at eye level I raised my hand to lift it out of the way... only to realize it was a seven foot (maybe eight foot) long snake."

What do softball, snakes, posts and El Pilar have to do with each other? El Pilar is one of the bigger Maya ruins in the vicinity; Pilar stands for pillar or post. It was to be included on the Eco-Tour being developed by CCDC. Problem was, we did not know where it was actually located!

I did know was that it was located north of Xunantunich, on the Guatemalan border, almost directly east of Tikal. It was somewhere past Bullet Tree Falls (BTF). After four visits to the Archaeology Department, in Belmopan, I learned it was ten kilometers from the river at BTF and "close" to the road. On Thursday, March 5th, I went with the guide being trained for the tour, and his friend Seed (named for his consumption of pumpkin seeds), on an exploratory trip in the "infamous" Toyota.

The three-mile portion to BTF was rough as usual. Past BTF the road actually improved for two miles. We started climbing a limestone ridge. It was rough going, tougher than to Caracol. There was one hill after another. I did not realize the elevation rose so high near San Ignacio. Halfway up one steep hill one tire went flat. The bald tires were giving out. We wedged huge rocks under the other three wheels (the hand brake is nonfunctional) and made a quick change.

The Belize Maya Site guidebook says to ask locals the way to the ruins... there were no locals to ask... just unattended *milpas* (farms). Figuring it was somewhere over six miles on the odometer (which actually works) we should have been sighting the ruins or at least mounds.

We were on a plateau, and this looked like a likely locale for a ceremonial site. Seed insisted it was still two kilometers ahead. Not satisfied that this was the case I was intrigued by two tree-shrouded mounds about 300 yards apart - about the same distance separating the Baking Pot main structures (a Maya site near the Spanish Lookout ferry crossing). I parked in the shade of some trees. Two Chicleros were also having a rest.

Seed borrowed their machete, and we followed a trail toward Guatemala. There were neither side trails to either hill nor any small mounds indicative of residential structures. Seed had the lead, but there was no need to "swing" the machete because Chiceleros, sneaking in from Guatemala, kept the trail clear.

There, intertwined on the branch ahead of me, was the thickest green vine I had ever seen - twelve inches away at eye level I raised my hand to lift it out of the way... only to realize it was a seven foot (maybe eight foot) snake. I did not have a measuring tape along.

SNAKE SLITHERS HITHER

It certainly was a fine specimen. I was reluctant to call attention to it as I thought Seed might try to cut it in half with the machete. Leo caught up to me and was not going to have anything to do with the snake. For all his years in the army, having been shot by marijuana growers and done bush trips, he was not taking chances. I ducked under to show that it was not harmful (what did I know?). Seed came back and assured Leo it was "no big deal". Seed gently nudged it with the machete urging it back into the forest where it disappeared. We continued along the path but could not find ways into the hills. We turned back. The resting Chicleros had gone. These Guatemalans on the Belize side of the border were not taking chances on us being Immigration Officers. Seed left the machete in a tree for their retrieval.

We continued along the road until spotting a logging trail leading into Guatemala. Forty yards along this turnoff there was a pathway headed into the thick brush. We investigated. Three dozen paces in was a small mound. Walking around it was an excavation revealed. It looked like a looter's trench - a narrow cut into the back of a burial mound, with a hole just big enough for a small person to crawl into a tomb. There was nothing discernible to the south or north, so we walked the logging trail.

There was a long mound parallel to the trail, possibly some kind of temple boundary. Again access was limited. Seed insisted there were ruins ahead. Travelling another hundred yards further, I detected another opening in the brush. This time we hit pay dirt. There was a huge steep-sided mound. As I climbed up, I noted large brick-like rock structures indicative of temple mounds. At the top, to the south, we could make out another larger mound.

Scampering over it we found an incredibly large surface area on top with several smaller mounds. Perhaps this had been a living area for royalty or priestly elite. We climbed around the edifices noting steep sides. Peering over these back edges, we could see huge chunks of excavation taken out of the sides. They were two meters wide, huge ugly intrusions into a majestic relic of the Maya past.

I mistakenly thought that these were archaeology excavations - the edges were smoothly and uniformly cut. This was El Pilar; it was bigger than either Cahal Pech or (visible portions of) Xunantunich. Those two were near rivers. This was in the mountains with no convenient water source yet it must have been an important site to be so large.

GETTING BACK

I was not anxious to return down the steep, bumpy road. Besides I was curious to see if we could find a "direct" route to Aguacate Lagoon - one of the attractions to be included on the proposed tour. We did find a route, but it involved crossing an unbridged creek, and then it tapered into an overgrown trail with fallen logs, thick underbrush and swampy portions. Distance wise, it might be a short cut but time-wise, it would be a delay. It may offer appeal to some adventurous travellers but may cause more consternation among tourists. I wanted to take the truck in for new tires. The CCDC Transportation Coordinator vetoed that, saying he could save $5 per tire by going to Spanish Lookout. It would have been less costly to have changed them in San Ignacio.

BULLET TREE FALLS WOMEN'S SOFTBALL TEAM

I have been coaching the BTF women's softball team, an excellent group of players with strong but wild arms. I had them concentrating on throwing for accuracy, speed to come later. On account of errors opposing teams were getting too many bases. They had a game against the Santa Familia team. It was four miles to the ball diamond. Rides took me a mile here, a mile there and shank's mare in between. The game was to start at 1:00 p.m. but did not commence until 3:30 p.m... Unable to get enough women together to form a team, half of Santa Familia's players were men.

They wanted me to umpire, but the rules are different from the softball played in Canada. Sure they threw underhand, but fast; base stealing was allowed; you could go out on a fall ball; and only nine players on the field. No, I was not going to umpire. Besides I wanted to record statistics by noting who made it how far around the bases and who caught how many fly balls etc.

BTF won 21-15. The catcher made a fantastic play at home plate, putting two out as they, seconds apart, tried to steal home. The pitcher, played solidly, accounting for the first three outs and getting the most runs. The captain was another excellent player, batting consistently and playing a strong center field.

The field behind first received a disproportionate amount of the action, with one outfielder putting out five by catching flies but also letting eight make home runs by missing bounces.

After the game, we went to Las Casitas resort, for a drink and a possible ride back. The CCDC Field Officer for Branch Mouth was there with the CCDC Nissan (in worse shape than the Toyota).

She had news that a brother of one of the players had two flats with the Toyota on the Pilar road while cutting posts for the Crystal Hatcher's project. I went on an unlikely sunset rescue mission with the Nissan.

Postscript *2001: El Pilar is now well marked with wooden signs pointing the way. Where Guatemalans used footpaths to pilfer buckets of Belize chicle, Belizeans were hauling out Guatemalan mahogany by the truckload on the logging path.*

Postscript *2019: The road is not well maintained and is a rough ride even for 4WD*

CARPENTRY SKILLS

Most of my place had been furnished with inexpensive furniture built by Belizeans: a spare bed, dresser, a night table, two square tables, and chairs. I was able to get a large waterbed at an auction in Spanish Lookout for less than a regular bed. I had not been able to find a bookcase.

At one of the woodworking shops, I requested a three-foot-high, two-foot wide six-inch deep bookcase to shelve pocket books. I was quoted 90 dollars. Ninety dollars! The small bed cost less than that; the dresser drawers cost less; my two tables cost less. For seventeen dollars' worth of mahogany, I made three bookcases to my specifications.

I borrowed a hammer and saw, bought some nails and proceeded to manufacture. Fortunately, the timber was the required width as my saw cuts were neither straight nor smooth. Mahogany is tough to penetrate. Two out of three nails bent going in; other nails stuck out every which way.

At least I lacquered on the undersides. One of the bookcases was a beautiful deep red. Any book would be proud to sit in that shelf. When I was robbed (*read "Robbed" in Chapter IV*), the only pieces of furniture they did not take were my bookcases. Was that a reflection on my carpentry skills?

TOO LATE

March 12, 1992

The bridge is long; because the river is wide, wide and rapid. I crossed the San Ignacio suspension bridge to find a teacher at the Santa Elena Elementary School, standing on the east side of the Macal River. Others were looking up and down the river.

Walking along the riverbank, I met a woman crying and holding a two-year-old. I asked her "What was the matter?" She said that her baby was missing. I quickly scanned the surrounding area. The river flowed quickly. "How old is your baby?" I asked.

"One year" she sobbed. The river could carry a one-year-old away easily. I sprinted along the path, parallel to the river, to see if there was a child at the edge, hoping for a sight of the toddler in the bushes or behind a tree.

A teacher and his class were coming the other way. I asked if he knew about the missing child. "Yes," he answered, that was why they were here looking. He advised me that there was no way the baby could be further downstream. Still, I proceeded as far as I could go, about two hundred fifty yards. Walking rapidly I noticed things floating by.

I tried to calculate how long the baby had been missing. If the teacher was here with his class, chances were that he was called before recess. Even if the baby had stopped breathing it might have twenty minutes before it drowned.

If the baby had fallen in twenty minutes ago, or more, it could be more than a mile down the Macal River. With the others, I turned back to search along the embankment. The more we looked, the more hopeless it seemed; it was a greater likelihood that the river had taken the baby. What else could one do? I started downstream again. The river is very wide at that point.

I offered to inform the police and hurried up to the bridge, across it to the police station. Several officers were standing around. My presence went unacknowledged. I accosted one, pleading for help to find the baby. "We know all about the baby" he haughtily informed me "a vehicle has been sent". Yes, I could see it winding its way along the side of the river opposite to where the baby fell in.

I wished I had thought to throw in a branch or log where the baby could have fallen, to see where it would drift. Fifty yards further downstream the baby was found drowned.

A TRIP TO GUATEMALA

March 1992

Dr. Pat Asling, CUSO Dentist, and I went to Guatemala, in her van, to study Spanish. Guatemala has beautiful scenery and durable cultural traditions. Some of the roads are well-paved highways. Some are bad dirt rut infested nightmares. We travelled a lot of the latter. The road down the east side of Guatemala was so bad we tried to return via Mexico.

The first day, out of Belize, went as expected - with 60 km of rough road before hitting the Flores-Tikal paved roadway. We stayed the night in Flores in a nice hotel for 50 Quetzals (5Q=$1US). Flores is on an island, and the many shops and restaurants carry on the Spanish motif. Unfortunately, new buildings are going up without retaining the old character.

What was even more unfortunate was that we had been under the impression that the road from Flores to the southern highway was a paved one. It was bumpier, more bone jarring and dustier than the section from the border. It took eight hours to travel 100 miles. There were quite a few times that the van bottomed out with a sickening thud.

Shortly after nightfall, we located another nice hotel at Doña Maria for 27Q ($5US). The rooms backed on to a walkway that runs parallel to a stream. There was a swing bridge across the stream that led to a park.

The next morning we visited the archaeological site of Quirigua. It is famed for its impressive stone stelae carvings. The site was difficult to find as there were no road signs indicating where to turn. It was located four kilometers into a huge banana plantation.

The stelae and other stone carvings were incredible. They were exquisitely carved in three-dimensional relief. One resembled a North West Coast totem pole in physical appearance (size and style). Even the colour scheme of using black, red and blue-green corresponded. Both totem poles and stelae give information on deceased leaders.

On the road down we had traversed through heavy forest that gave way to a dry area filled with cacti. As we arrived in Guatemala City (pop. two million) we could discern volcanoes bordering the southern edges. We were advised to study at the Francisco Marroquin Spanish School. It took an hour to reach there.

Antigua is a world heritage site. It was a former capital (from around AD 1545 to 1755) until an earthquake destroyed most of the beautiful churches. There are about twenty churches in various states of reconstruction or ruin. The town is situated at the base of Mount Agua, where people make eight-hour pilgrimages to the top.

Several other volcanoes are nearby. Flowers blossomed on the trees, and there seemed to be a parade every second day. Pat and I signed up for seven hours a day of Spanish instruction - verbs, verbs, verbs. We were each assigned to live with a local family.

During breaks from the one-on-one language instruction children, in colourful Maya dress, accosted us to buy handicrafts. In the evening there were charming restaurants to attend or bars to frequent. The place was full of visitors, mostly a casual, almost bohemian, type of tourist - adventurer types. Rich Guatemalans came to Antigua on weekends. Things were inexpensive: tea or coffee 1 Q (twenty cents) and 20 centavos to mail a postcard (four cents).

On a weekend off, we went to Lake Atitlan, surrounded by another grouping of volcanoes. Again more crafts for sale and more bad roads to reconnoiter. We had been told that the Sunday market in Chichicastenanengo was spectacular. Unfortunately, other visitors, to Guatemala, had heard the same. We could not locate a hotel room. Even at 500 Q they were full. An enterprising villager offered a garage with beds for 50 Q, that we accepted.

The market was colourful and immense but just a huge version of the other markets. It was a bit depressing - there were all these people selling things and so few buying. Most of the tourists were taking photographs and not making purchases.

We headed back along the windy road and stopped at a lookout where a family handicraft project was on display. We politely said no to bargains. I started eating cornflakes out of the bag. One of the women put her hand out. I gave her some. A girl came up, and I gave her a handful. Then they all lined up for handfuls. After that, they asked for their babies and sisters.

I had been carefully distributing to each in turn when Pat said: "Why don't you give them the whole thing?" They were Pat's cornflakes. I handed the bag over, and there was a big scuffle. Things are pretty bad when five women, of the same family, fight over cornflakes. We did not have extra funds to help them further.

For a couple of nights, I had severe chills. I would put on three shirts, two pairs of pants and five or six blankets. Then after freezing for a couple of hours, I would break into a sweat, and everything would come off.

In the daytime, the sun seemed to bother my eyes even though I had photo-gray glasses. After I took malaria tablets the bouts of fevers and chills stopped.

There were two more days of Spanish for me and then I spent one day in Guatemala City researching institutes for tips on tanning animal hides. In Cayo, I am helping a local group with their leather project. So far I have convinced them to go with organic tanning (rather than using toxic chemicals). They are hoping to use mahogany bark but cannot find information on its suitability.

There is a research institution in Guatemala City, but it no longer had a tanning expert on staff. They sent me to the forestry department who could not help, but I did obtain the name of a reforestation expert.

During my walks, I was able to visit five museums: two archaeological, two arts and one natural history museum. The zoo was also impressive with specimens of grizzlies, lions, jaguars and many other animals. There did not seem to be a lack of funds for cultural attractions.

After two more days of Spanish, we took a trip to the Quetzalas Biosphere. The Quetzal bird is Guatemala's national bird. It is the most resplendent bird of the New World with two-foot-long tail feathers. Demand for these feathers has led to its being a rare bird. If you get up at 6:00 a.m. you might get a glimpse of one of the 600 pairs flying high through the treetops. I slept until 10 a.m.

We thought the road too terrible to return the way we came. We tried to head through Mexico. We overnighted in Huehuetenango before confronting the Mexico border. We crossed at a place called La Mesilla. They did not have a supply of the stickers needed to allow transit through Mexico. We were told to go to Cuidad Hidalgo. We thought it a few minutes away.

It turned out to be four hours away, on the coast in the wrong direction. After several futile attempts, we gave up trying to explain we had already crossed into Mexico way inland. We had to photocopy all documents twice (and find a place to do that on our own) and then drive 30 km to the border check to find out that it was not sufficient.

We needed Mexican insurance not obtainable on a Sunday. We stayed overnight. In the morning we had no luck. The agent wanted $400 US for two day's insurance or a $60,000 deposit (the value of the vehicle). Back through Guatemala - except it took five hours to get the necessary documents and wait through the line-ups.

We proceeded along the coastal highway that would dwindle into a one-way street, usually with us going the wrong way. We would hit and miss our way out of towns. One place took an hour. The next morning we retraced our routing along the Southern Highway - which facilitates overland shipping from the Pacific Ocean to the Caribbean Sea.

There were only a few trucks at 5:00 a.m. We made good time until we hit the bad road to the north. It was twenty kilometers per hour time again, sometimes five, sometimes 0.5 km an hour.

We bottomed out one time too many. A clackety noise emanated from the oil case. Pat stayed with her van while I hitched a ride back the way we came for thirty kilometers. Each time the transport hit a bump (about once a second), it felt like being punched in the stomach or the kidneys. After I located a helpful individual, it took five hours to secure a tow back, which made for another overnight. Fortunately, the problem was fixed on the third try. We only lost 23 hours due to my driving five kph in a two kph zone.

NORMAN AND RICHARD'S
ALMOST EXCELLENT ADVENTURE

April 1992

Two friends arrived from Calgary to attend the *First World Congress on Tourism and the Environment*. There were two days of sessions and two days of tours and discussions.

After meeting them in Belize City I "endured" a round of drinking during the cultural evening hosted by a pay-for-yourself bar by the Ministry of Tourism. The dancing and performances were great. The Garifuna were particularly good, and there was an excellent modern dance decrying logging, by a Belize City troupe. The Garcia Sisters performed a traditional Maya dance, in stately unison. There was a spectrum of presentations.

A few wanted to redefine Eco-Tourism to ensure the environment received something from the visitations. Some cried that government regulations and having to hire locals were ruining the business. A representative of American Investors stated, "it was a `Right' to go anywhere in the world to make money - only profit earning could save the environment."

From the "enlightening" presentations participants went to different districts to enjoy various aspects of Belize. They divided into discussion groups on special topics ranging from starting up an enterprise to community participation. Norman went south to Punta Gorda; Richard came to San Ignacio to visit some local projects. At the office, I worked on a reforestation proposal.

We met at my place for the weekend. It was difficult to induce Richard and Norman out of my backyard as they noted all kinds of birds flying around my residence. They are avid bird watchers (are there any other kind?). There were chickadees, wrens, blackbirds, hummingbirds, woodpeckers, and a dozen others I was unable to differentiate.

Enticing them with promises of richer finds - more diverse bird offerings - we headed up the hill to Cahal Pech Maya site. Along the way, they did locate many species of birds. It was slow progress. I should have just pointed out the way. What usually takes an hour took all morning. They might have spent all day if I had not been along.

We sought out Aguacate Lagoon. When we passed Baking Pot, the Maya site near the ferry, I only pointed it out. I knew there were lots of birds in there. They enjoyed the quaintness of the hand-cranked ferry before we cruised through Spanish Lookout - or rather tried to. From the ferry to Aguacate it was "stop, stop, stop, back-up..." to see... more birds... parrots, flycatchers, vultures...

Fortunately (for me), there were not too many birds at Aguacate Park (only enough to stay for two hours). Aguacate has several features: the lagoon, a Turtle Puddle, Reptile Pool, Green Hole (a sinkhole of vegetation), White Robe (which Richard with the Ph.D. in Geology says is not a natural placement) and a Maya site. We left before dark and arrived at Las Casitas for some sunset bird watching. This is where the Mopan and Macal rivers meet to form the Belize River.

There were iguanas as well as birds to see. They were excited, if I remember correctly, about spotting a swallow-tailed kite. At last the sun set and I figured that would do it for bird watching. Oh no! They spotted some herons, egrets, and ibex in a swamp we passed.

RAFT RIDE

The next morning we tried for the Mountain Pine Ridge area (caves, falls, pools, maybe some birds) but the CCDC vehicle was acting up. At Monkey Falls the 4WD stopped dead. Norman rigged up an improvised throttle mechanism. The push starting and the engine's coughing were too much for Richard, so we went to Clarrisa Falls. The plan was to have lunch and then leave them at Xunantunich, a few more miles west. Chena, the Manager, told them how they could raft down the river and watch birds along the way. They were so captivated with this idea that I left them and went to coach softball.

The Bullet Tree Falls Comets won 26 to 7. I was home by 4:00 p.m., a bit early for them to be back from rafting. At 7:00 p.m. I became worried and went looking for them. In case they came while I was gone, I left a note telling them to meet at the Red Rooster for pizza. They were not at Clarissa Falls. Neither had Chena returned from picking them up; she was overdue. She left at 4:30 p.m.; it was now 8:30 p.m. At Bullet Tree Falls where the pick-up was supposed to be, I found Chena making phone calls from the village phone. There was no sign of Richard and Norman.

Two brothers from Miami, who had tubed the river ahead of the "birders," were posted as lookouts along the east side of the river. I headed up the opposite side, waking up people to ask if they had seen two gringos rafting. The last reported sighting of them they were going through Calla Creek. "They were going slowly, watching birds" was the description given. That was them all right. I had visions of them overturning and being marooned on some bank or having pulled ashore on account of it being dark. By 10:30 p.m. there was still no sign of them.

Chena and I planned to re-raft the river, which she estimated would take five hours at night. I was thinking "oh great! I have to spend five hours paddling at night because those wimps wouldn't.

To get back to Clarissa Falls we had to pass through San Ignacio. We checked at the Red Rooster. Norman and Richard were there eating pizza and drinking beer. They had landed at 7:00 p.m. They had walked the path into Bullet Tree Falls and hitched a ride.

Chena would not talk to me for about a year.

A MOVE

July 1992

The owner of Maya Mountain Lodge was looking for some to stay at the sixty-acre estate across the road. The main house is a two-story, four-bedroom mansion with bathrooms upstairs and downstairs. There was a work shed, storage shed, double carport, two large cement water tanks, and huge lovely trees, providing shade, including coconut-bearing palms. In the back was a large pond, pineapple orchard, cohune palms and no neighbours visible in any direction. I moved in that afternoon. I had to sweep up dead bees (where did they come from?). While cleaning in the bathroom, I heard a shaking noise from the cardboard box of cables and cords I had pushed into the corner. This nasty looking snake appeared! It rose to about chest level and swayed back and forth. It hissed and spat at me each time I came into the bathroom, looking me in the eye with a wide open mouth as if to say "I'm going to swallow you". It was a thunder and lightning snake (not dangerous to humans) but very territorial.

120,000 DEAD

August, 1992

At my new place, there are lots of parrots. I could hear an owl. Definitely woodpeckers and herons were out at `my' pond. Hundreds of bullfrogs came to the house at night to feed on the bugs attracted by the outdoor lights. Horses, from Maya Mountain Lodge, wandered freely about.

There was carnage everywhere. After returning to my new home, where I had spent the weekend cleaning up, I found dead bees piled ankle deep. The walls had been ripped out to get at the hives. The Agriculture Department had located three colonies, of 40,000 to 50,000 bees each, and the onslaught began. I thought there might have been a more peaceful solution such as just moving the combs of honey to a different location. The bullfrogs had a feast picking off the straggler bees as they vainly searched for the colony after returning from foraging in the flowers.

So far no more snakes living in the bathroom. It has been cooler here. Warm but not uncomfortably hot like April, which is surprising because the sun is now directly overhead at noon. The sky is cloudy, off and on, which probably keeps the temperature down along with bouts of rain now and then. Prime hurricane season starts soon.

WHEN I FIRST MET MANFRED

August 1992

When I first met Manfred, I thought he was one of the local Amish because he would drive a horse-drawn buggy, had a beard and wore a straw hat. He, in turn, thought I was a local Mennonite because I had a beard and wore suspenders.

Manfred is from Germany, and his wife is from England. They came to Belize over a decade ago. Their seven children aged 16, 14, 11, 8, 6, 4, and 2. Living a mile down the road, they are one of my neighbours. Twice a week I go there to buy milk and avocados. They live off the land without electricity; their drinking water is collected from the roof.

Manfred is Chairman of the local Macal Dairy Cooperative. He delivers milk to town every morning and sometimes stops in for coffee on his way back. In the mornings I try to work at home because of the pleasant surroundings. One morning he asked why I was a vegetarian. I responded, "Why aren't you a vegetarian?" He had been a vegetarian after being in India. That lasted until he was left stranded on a deserted island.

In 1971 he had hitched a ride on a yacht with another German, his English wife, and their thirteen-year-old daughter. He joined them at the Panama Canal and was heading to Peru, to see Machu Pichu, and sailed with them as far as the Galapagos. Along the way, they stopped at an uninhabited island, Cocos Island, south of Costa Rica, not far from Columbia. For about a week they tried living off the land. The Captain did not live off the land too much.

He had 600 bottles of rum as ballast that decreased by a bottle a day. In the mornings no one could go near him because his hangovers made him unapproachable. In the afternoons he was incoherent.

Disagreements arose. Captain Rum was indignant when Manfred would not obey orders. He demanded to know who ate the last of the sugar and cinnamon. The Captain's pride and joy, his daughter, said it was Manfred. Manfred said it was the daughter (which I'm inclined to accept, as Manfred does not take sugar in his coffee). The yacht left without Manfred, who was told he would be shot like a wet rat if he tried to reach the departing vessel.

A month later another sailboat came by. These people took on freshwater and palm leaves for bedding. They offered to take Manfred with them. Their destination was Australia. They had to stop for provisions because they had just done a dope deal in Columbia. The guy who had sold them the stuff informed the police on them (so that he could get the dope back).

That is why they had left Columbia without provisions - they were under a blaze of bullets. Manfred opted to stay on the island.

On the island, Manfred existed mostly on coconuts (which explains why he turns down offers of coconut) and whatever animals he could capture: scrawny seabirds and one wild boar. There were a number of wild boars on the island. He dug a pit and placed stakes at the bottom. Thatching and broken coconuts were sprinkled over it to lure a boar. Sharing pieces of coconut with one of the boars, his only friend on the island, he discovered that they like coconut.

Unfortunately, the only boar that would go near his scent was his friend. That was the one that fell into the trap.

Two months later another yacht arrived. This was a huge one owned by a rich, stingy Californian. Some of the students crewing told Manfred that he measured their cornflake ration in the mornings. He would not give Manfred a ride claiming that there must have been some reason Manfred was left there.

When the crew protested, he pretended to radio to the German Embassy in Costa Rica. The yacht left. A rescue did not materialize.

Manfred was left with two tins of soup and a book. The book was about how people are not satisfied with their lot in life; how people say if only they could make a bit more money; if only they could get married; if only they could buy a house ..." and of course if they did accomplish those things they would be looking for something else. Manfred had decided to live with his lot (he was not going to wish "if only I could get off this island"). After I finished laughing at his misfortune, Manfred told me that he had marked, in his shelter, with a knife "Try not to laugh too much about it after". Another month and a half later another boat showed up. This one was the Beta II, a treasure hunting catamaran with a diving submarine.

They were not too happy about Manfred being on the island. They were looking for treasure. The island is rumoured to be a stopping off place from buccaneer raids on Spanish gold shipments, coming north after looting the Inca Empire.

A Canadian was there in the 1800s. He returned to the Maritimes and built a mansion, without saying where he secured the financing.

Manfred does not know if the crew of Beta II discovered a treasure, but they offered to take him off the island in return for the rights to his story. Manfred thought it was a fair deal. When he returned to Germany, a major magazine ran a six-page article on his ordeal.

PUPPIES

October 29, 1992

After moving into the "mansion", I had been broken into and robbed. The police recovered most of the items. Manfred has bought the place and plans to move in at year's end. In order to reduce the chance of further break-ins (my hand-painted sign of "Beware of the Snakes" was not effective against illiterate trespassers; although it was a favourite of tourists). Manfred offered the loan of one of his dogs. Tobie was pregnant; I thought I could find homes for the pups.

Tobie had her puppies. She would not eat anything this evening - which was peculiar, as usually there was no end to her appetite. She was restive, so I moved her around to the garage side of the house; there she eagerly headed into the "guard/generator" house. Looking like she had picked a spot, I cleaned away the debris and swept up what I could.

Every half-hour I checked on her. At 9:00 p.m. still no pups. I figured it could still be days away. At 9:30 p.m. there were six puppies: white ones, brown ones and black ones. There was a runt, and they were all suckling. With eyes closed and hind legs pushing they each managed to find a teat.

At 10:30 p.m. there were eight puppies. Momma had moved into the dirtiest part of the guardhouse. What was the attraction of the greasy, oily dirt? Earlier on she had let me touch the puppies, but now she growled when I came close. To bed I went.

At 3:30 a.m. I heard some clattering noise and some growling; I investigated. Nothing unusual around the yard but now there were eleven puppies; they were sneaking in from somewhere. I tried to entice Tobie into a more comfortable spot, but nothing doing; she liked the dirty corner and licked her litter cleaner.

By 4:30 a.m. Tobie was eating anything. It was the first time she was away from her pups. She returned to give each a turn at feeding. Whether this was by design or by accident, I'm not sure. Sometimes when she licked them they lost their place at the feeding station; other times when she got up, they started jockeying for position all over again.

How they managed, to find their way by random direction was amazing. They took an inordinate length of time to make a micron of distance, with back legs churning perpetually and front limbs loose fish flippers. But manage they did.

I do not know how the mother managed jockeying eleven, each of them with their own whine. Soon Tobie was out walking and eating ...

I'd see her out by the road; she needed a break. On one of those outing, I moved the puppies into a big box; the runt was dead. Tobie returned and would not let me take it away. I placed it on the ground pronouncing "it is dead". She carried it in her mouth, took it to the dirty spot, lay down with and licked it. A battle of wits and subterfuge ensued before I could get it away and buried.

Postscript: *The puppies turned into fine dogs; homes were found for all remaining ten.*

HALLOWE'EN

October 30, 1992

There were a number of Hallowe'en parties scheduled for the weekend; I thought it would be interesting to see how the custom was locally celebrated. On the Friday night, Caesar's a place on the Western Highway about ten miles east from San Ignacio, held a party. I didn't know there were so many "white folks" in the area. Some Creoles and a few Mestizos showed up but not in costume. It seemed to be a "gringo" dress-up occasion. It was a good party, good music, good costumes, and good food - and then there was Mervin.

When I was introduced to Mervin, the first thing he said was, "I'm looking for a wife". Not "How do you do?" nor "glad to meet you" nor "hi." Before I could formulate a reply, he was off on, "I planted four acres in hardwoods, yep four whole acres." It was noisy; I lost the rest of his story. There might have been a connection between his statements.

One of the fellows at the party, a chiropractor, stationed at Belmopan, was originally from Missouri. He told me about some of the art deals he had tried to inaugurate in Belize in the seventies. His eight-year-old daughter Beth, who had dressed up as a clown, was crying. She was worried about not winning the costume prize because she had washed off some of her make-up. Her father tried to kid her "Are you worried about the prize money you will lose?"

"Noooo!" she wailed.

"Well then it must be the adoration of millions" he taunted further.

Missouri left to go for a walk, so I asked her "Who do you think has a better costume than you?"

"The girl in the princess outfit" she sniffled.

"Yes," I replied "that is a nice costume but not as good as your costume. Who else?"

"The girl in the Raggedy-Ann costume" she offered.

"Yes," I affirmed, "that one is almost as good as yours" (this entrant won third prize).

"But I'm not going to win because I took off my make-up" she cried.

"Don't worry about that" I consoled "it still looks like you have some on. Besides your costume is good enough without it. Stick around for the contest." She stopped crying, and when the contest was held, she took first place. She was one smiling clown!

Sometime later, while I was sitting with some others, Mervin plunked himself down in an adjacent chair and slapped his open palm on the table and said (to no one in particular). "I bet one hundred dollars that no one can refute any of these three claims." He listed them:

1) "There was no syphilis in the New World before Columbus;
2) There was no malaria before Columbus; and
3) The Spanish taught the Indians to make tortillas."

Some people took him up and discussed the merits but quickly lost interest. He then wanted to know if anyone knew what "yaws" was?

I said it was what an airplane does when it's not "pitching".

He said, "it was a second cousin to syphilis".

Perhaps because I was sitting beside him, he started his next barrage at me. "I have an 8,000 BC pot from Thailand," he exclaimed, "it's not worth anything to anybody."

Me: "Maybe it's worth something to Thailand?"

MERVIN: "It ain't worth nothing to no one."

Me: "Then why do you have it?"

MERVIN: "I didn't steal it or anything, I got it in Thailand. I've got Maya ruins on my property, and I don't touch them. There's walls this high" (he gestured) "and temples and structures, but I don't touch them ... I tell my neighbours, the Mennonites, that they should not touch them."

Me: "That's the best policy."

MERVIN: "Do you know where I live?" he demanded and then answered "I live near Upper Barton Creek. Do you know where that is?" and without waiting for a reply, "I'm two miles from there."

Me: "Which side?"

MERVIN: "What do you want to know for? Is this some kind of investigation? I bust heads you know. Anybody that comes messing with me I bust their heads. You know a year back some blacks come and took the roof off my neighbours. I bust their heads and then there was peace for a year. People say I am bad, but I bust heads and then there is peace. My neighbours [Mennonites] don't believe in that, but people take advantage. So once a year I bust heads. Sometimes I win, sometimes I take a beating."

Me: "That's effective, is it?"

MERVIN: "Let me tell you. This Guatemalan, he tried to slice the top of my head off. He got me across the eye, then he hacked my bone in my elbow, but I backed off, and to this day he don't come on my property. No sir he goes the long way around!"

Me:"That's quite the technique you have there."

MERVIN:"About every year I bust heads and then things are quiet." *Pause* "What do you do in the states?"

Me:"I'm not from the states."

MERVIN:"What do you do in the states?"

TERRY:"I'm not from the states."

MERVIN:"What is it you do from wherever you are from?"

Me:"I'm in Cayo, and I'm involved with Economic and Community Development."

MERVIN:"Bullshit"

Me:"Well f___ you too."

MERVIN:A pause "Hardest job in the world." [I don't know if that was in response to my last retort or my explanation of what I did.]

MERVIN:"How long you been doing that?"

Me:"Mervin, our conversation is at an end."

MERVIN:"No, how long you been doing that?"

I left. You meet spooky people at Hallowe'en.

Postscripts*: When I saw Mervin after that either he did not recognize me or was not about to strike up another conversation. As far as I know, he still has not found a wife.*

Two years later the chiropractor was in a nasty automobile accident and no longer practiced.

THE BELIZE ZOO

The Belize Zoo is located halfway between Belize City and Belmopan, the capital. It had an impressive set-up. Their focus is on taking animals that had been hurt or young found without mothers. Mammals, birds, and reptiles were caged in environmentally familiar surroundings in an attempt to replicate their natural habitat as much as possible. Not only did you see the animal and the vegetation it dwelled in but both of them together. It was a holistic approach. Moreover, the animals behaved naturally. One had a sense of being in the rainforest observing them, as from behind some tree. On occasion, one had to search a little to pick them out - just as if you were seeking them in nature. When one looked closely and carefully jaguars could be seen lying in the shade of a tree just like they would do during the day out in the jungle. The designer of the zoo has received international awards for the unique approach.

A DAY IN THE LIFE

CCDC no longer works with the 21 de Septiembre Grupo or the Las Vegas Group. They have taken on the Western Leather Cooperative and Teakettle Survival Group. The former plans to organically tan hides to make leather products; the latter group plans to raise pigs.

The day started with taking members of the leather group to the area behind Cahal Pech, to collect tanning bark. An instructor from Guatemala showed them finer points of tanning, which varied somewhat from the vertical tanning I promoted using mahogany bark (from cut timber). He was familiar with tying skins into a ball and dipping them in solutions of crabbo bark.

We were stripping bark from half the side of tree trunks (so that they will not die) on one of the member's property. I was amazed at the number of ancient earth covered structures dotting the hilltops of the land. He, in turn, was amazed at how quickly I had been able to pick out the features. We stripped enough bark for the first tanning, leaving bark on the trees so that they would live on.

After that, I went to see how the technician, from Florida, was doing with installing the incubator hatchery equipment for the chick producing project. When I had supper with him, he said he spent a lot of time in Calgary installing ostrich egg incubators. From there we drove to meet with the Branch Mouth Women's Group. We moved the plucker from one member's yard to another. It is amazing how it removes feathers in minutes. It takes hours by hand.

CYCLING ON THE JOB

On a couple of recent field trips to help complete the work I took along my mountain bike. Last Tuesday, Ali Flores (a new CCDC Field Officer) and I went to San Antonio, eight miles south through the beginning of the Maya Mountains. The group there has a large pig raising project (funded by Canada Fund) and peanut crop. It was time for the year-end evaluation.

When we arrived, we found there was a commotion over the death of one of the pigs. We called the vet. Ali weighed the weaners and castrated the ones eligible for that incisive operation. I conducted the evaluation, with the group's President, then cycled back to Cayo. It was a rough road but mostly downhill. Devil's hill was the most difficult, but I arrived back as it started to rain.

On Saturday Ali had a meeting with the Progressive Group of Santa Familia. The Progressive group grew corn and soya; they planned to establish a feed mill to compete with the Mennonites. Although only two miles away as the crow flies it is seven miles by road. Juan had booked the truck that day to take the Teakettle group to visit bio-gas operations. Ali took the motorbike, and I cycled there.

It took just over an hour to do the seven miles with three stops for soft drinks. We were both there at 2:00 p.m. but the group thought the meeting was at 3:00 p.m. There was no rain that day and no cloud cover. After my part of conducting the evaluation, I cycled back. It was another hour with another three stops for soft drinks. The hardest parts were the hills between San Ignacio and Bullet Tree Falls.

Early Sunday morning the Coordinator picked me up to meet with the Los Tambos group to do the evaluations of their pig and chicken rearing projects. It took three hours to reach there and round up the members from Santa Elena, San Marcos, and Los Tambos.

As the members of this group were almost exclusively Spanish speaking, I stumbled through the Field Officer and CCDC part of the evaluation, with the Coordinator out of the room, by using a combination of words and gestures.

He had a collective pig-raising project to discuss with them, and I had to compile and write-up the results. I cycled back the seventeen miles. Being

Sunday, Spanish Lookout, the Mennonite community, was shut down so I could not have any refreshment until Central Farm - twelve miles away. I had two soft drinks there, two more at the village of Esperanza and three more when I arrived home.

Thankfully the trip in from Los Tambos was mostly flat and downhill. After two hours I was putting the results onto a computer spreadsheet.

Postscript: *Ali Flores was murdered one night, after meeting a stranger in a Cayo restaurant in 2008.*

PLAYING FRANCIS FORD COPPOLA'S PIANO

Besides having accomplished two things seldom done in film (making a movie better than the book - <u>The Godfather</u>; and making a sequel better than the original - <u>Godfather II</u>) Francis Ford Coppola has a lodge in the Mountain Pine Ridge area, just down the road from me. He comes several times a year, most recently in April, with cast and crews from his movies. The lodge has a piano; when I dropped by I was invited to play.

I had gone to Blancaneaux Lodge to visit the Chuc's, the family I had originally stayed with when I arrived in Belize. They were caretaking at the lodge. Their daughter was a receptionist there, and their son was a waiter. The day I happened by there were no guests and only the receptionist, the cook, a guard, and a waiter.

As the guard was showing me around, he pointed to the piano saying that they had just gotten a new organ and did I want to play it? I looked at the piano, and I looked at him and thought, "If he does not know the difference between a piano and an organ, maybe he won't know the difference between someone who can play and someone who just makes noise."

I sat down to play the seven different versions of Heart and Soul that I knew. As I cannot even play top and bottom at the same time, they thought that was two different songs. The staff gathered around; they thought this was great. I showed one how to play the song with her left hand, and we played together; that sounded like another song.

VulCANOEING

November 19, 1992

November 19 is a national holiday in Belize. It is also the fourteenth birthday of Manfred and Janet's oldest boy, Johnathan. I took him and his brother Paul, canoeing up the Macal River, which was high from recent rains. We went seven miles upstream to the Chaa Creek Resort.

When I had made this trip previously, I had been able to start from the suspension bridge. This time we had to pick up the canoe from Cosmos Campground, another mile downstream. It was hard paddling just to get to the bridge because the current was strong. After that, we had almost two miles of calmer water.

The first set of rapids was not too difficult to pass. It took a while, but we made it through without having to exit the canoe. Tourists were paddling the other way (having paid to be trucked upstream) and applauded our efforts. The brothers were amazing finding iguanas high in the trees and birds hidden along the shores. They could do that and still paddle. I could only do one or the other. And when I did stop paddling it took quite a while to spot the creatures.

The next set of rapids was impossible to negotiate. When I had done it before, the river level had been low enough to walk the canoe through. Fortunately, these rapids, known as Monkey Falls, were near Devil's Hill where the oldest brother often went swimming and fishing. He knew of a side cut to get us around the rapids. When I got out to push, the current took one of my sandals.

Shortly before getting to Chaa Creek there was a heavy current too deep to wade, and we remained static. No matter how hard we paddled, the same trees and rocks stayed alongside us. We diverted to the right bank, losing only a few yards, where we were able to pull ourselves along by the branches of overhanging trees.

After four and half-hours of struggle, we arrived at Chaa Creek where we only had time and funds for some ice water. Soft drinks are three times more expensive than in town. We appreciated the current being with us on the way back. The return only took one and one half-hours.

Postscripts 2019: *There used to be eight outfits conducting canoe tours or rentals up the Macal River. All but one stopped after the dam was in operation, and even for them, it was just a part-time endeavour.*

Johnathan had been involved with protesting the building of the second dam. He became immersed with real estate and is also the owner of the Shell Station in Santa Elena and Cassia Resort (formerly Windy Hills Cottages). Paul is in charge of the lumber company he and Johnathan had started after success in real estate.

CHRISTMAS IN GUATEMALA

December 1992

For Christmas, I went to Guatemala and stayed with friends who worked at the Canadian Embassy. The day before Christmas they took us to tour a market in Guatemala that specialized in Christmas decorations. Craft persons were building nativity scenes out of straw and thatch.

Christmas trees were also under construction as the cutting of evergreens is not permitted. There were some black market ones available, but they did not look as good as the ones made from twigs and cuttings stapled to a pole. I wondered where the poles and cuttings came from; just how much forest was spared? Christmas Eve was a very noisy affair. Where we revere quiet for the occasion, Guatemalans celebrate with hours of firecrackers.

ANTIGUA

After Boxing Day I returned to Antigua to improve my Spanish. For $35 US a week, which included three meals a day, I stayed with a local family. My Spanish course was four hours a day, four days a week for $39 US. One-on-one instruction was from 2:00 p.m. to 6:00 p.m. leaving me the morning to study and also enjoy walks through the old capital of Guatemala, now a World Heritage Site. I managed to visit some tanneries. In Antigua, there is an enormous factory employing thousands. It is the only place in Guatemala that makes thread. It is made from locally grown cotton and sent to El Salvador to be made into cloth, then exported back to make the colourful items for sale.

IGUALEMENTE

In Spanish, there are many sayings and expressions to convey good wishes and well-meaning. For someone learning the language, it is not easy to discern neither how the expression translates nor what the appropriate response should be. There is a simple rejoinder "*Igualemente*" meaning literally "equally" or same to you. On my first day in Antigua, I was surrounded by many *muchachos* (young lads) trying to sell me a hat. As there would be exposure to lots of sun I valued a colourful straw sombrero and spiritually entered into negotiations with one aggressive "salesman". He started at sixty Quetzalas ($12US), and I offered half of half (fifteen Quetzalas). We settled on thirty. After the other boys had given up, a smaller fellow persisted that I should purchase a second hat. He followed me most of the afternoon, waiting for me to come out of shops, dropping the price of the second hat he was sure I would purchase from him. After refusing to buy, even at ten Quetzalas, he had an American expression for me "Phuck you, Senor." To which I had the rejoinder -"Igualemente."

Mount Agua

On New Year's Day, I climbed Mount Agua. It is a prominent dormant volcano rising to 12,000 feet as part of the backdrop to Antigua. The climb starts in the village of Santa Marie, ten kilometers from Antigua. I started out at 7:00 a.m. thinking I could catch a bus or hitch a ride.

Being a holiday, there were no buses on January 1st, and the only vehicles I noted were taxis coming from Santa Marie. As I walked along the road three villagers accompanied me; they had finished their night shifts in Guatemala City. They showed me short cuts, which were so steep that by the time I reached Santa Marie I was already tired. It did not help that I had been dancing until 4:00 a.m. at a New Year's Eve Fiesta.

By 10:00 a.m. I was slowly plodding along the trail. The upper half of the mountain was covered in cloud, but I hoped it might clear by the time I reached the top to see the Pacific Ocean. About halfway I was able to see Guatemala City and environs including the active volcano - Pacaya.

The rest of the way I was enclosed in a cloud and was unable to see more than a few meters. About four-fifths of the way I encountered three more people who had given up and were on their way down. All the other tourists on the mountain were on a package tour from Quebec; their trip that commenced in the Yucatan included most of Guatemala and finished up back in Mexico.

The cloud cover did not let up, so I did not get a view from the top. On the way down I hiked with a couple from Quebec. About a mile before the start of the trail I noticed another woman turning down a different path. I asked them if she was with their group, but she went around a corner before they could get a glimpse.

I did not recognize her as any of the ones I had passed going up, so although she was obviously a tourist, from the blue ski jacket she wore, she could have come up separately. The village dog had followed her. One of the Quebeckers said he should have realized that she was with their group. We were extremely tired from the strenuous climb.

In Santa Marie, New Year's Day celebrations were in full swing. Firecrackers and rockets were continually being set off. Statues from the church were placed on a large platform, and about forty men from the village carried it through the village in a rhythmic three steps forward two steps back fashion. A marimba band played, and soldiers were out in full force. I suppose a background of fireworks would be suitable for continuing the revolution.

The woman I had seen disappear down the other path had been with their group. Their guide was quite annoyed with me that I had not done anything about her going the wrong way. She did not seem to understand that I did not know who was in their group or what trails led where. At least I had been able to advise that I had seen her when she had not shown up. She showed up shortly, but two who went looking for her were gone for hours.

PACAYA VOLCANO

In Antigua, I signed up for the advertised "experienced" guided hike up Pacaya Volcano. They take people every day. Some thirty people had signed up to go that day, and it necessitated acquiring another van. The regular van was in bad shape; it had perfectly bald tires. The other van was even more decrepit.

As I was first on the list (which is unusual for a person whose last name starts with a V), I was called first and was permitted in the "good" van first. This meant I could choose where to sit, which was behind the driver so I could stretch my legs out between the bucket seats. This was to become important later on. The "guides" were actually organized enough to divide us up into two groups and assigned vans. That was the last bit of organization the two drivers, and two guides demonstrated that day.

After taking forty minutes to get everyone in, we headed to the gas station around the corner to take on three gallons of gas in each van. The first van went very slowly up the hill, out of Antigua. It had to stop frequently to wait for the "back-up" van. We took a turn that the other van did not. It was either a short cut the second driver did not know of, or the first driver did not know what he was doing (I vote for the latter). Two hours and thirty minutes later (on paved roads) we pulled into a store to buy supplies and stretch our legs. The place we were going to visit is only thirty miles from Antigua.

After the stop, we were off like a "herd of turtles" to the turnoff to the volcano. We stopped a few hundred yards up a dirty, dusty road to await the other van. When the other van caught up, its passengers were ordered out. Some were made to sit on the roof of the number one van, others were crammed in beside us, and three were made to hang out the open side door with only loose grip holds to keep them in place.

At first, the three American university students hanging on the edge of the van found it amusing and adventurous, but after five minutes they could no longer keep a grasp. The two on either side of the door had the vertical edges to hang on to; the one in the middle was having the most difficulty. His fingers slowly slid from the underside of the roof, as we rounded a bend; I grabbed his arm and held him in. We maintained steady progress up the mountain until the van overheated.

The driver jumped out with neither placing it in gear nor putting on the brake. The van started to roll back. I've had vehicles like this and recognized the symptoms of a vehicle that was not going to stay in place. I moved my outstretched foot to the brake and moved into the driver's seat before the other passengers realized we had been in danger of backing off the cliff. The driver had disappeared. We did not know if we were being set up for *banditos* or if the driver went for water. Some placed rocks behind the wheels.

While waiting around, our second stretch, we learned that the other van had a flat tire. David, from Detroit, who produces television commercials for Canadian Tire and Leon's Furniture, hitched a ride down the road with a truck descending the mountain. He had had enough. Tony, Director of Fitness at San Antonio University in Texas, and I start walking up the road. Others followed. Forty minutes later the two vans caught up with us, and we were "requested" to ride.

It turned out to be quite a bit further; we passed through a number of villages as we slowly made our way up the mountain. My Spanish was paying off as I could interpret what the "experienced" driver and guide asked in the villages we passed. "Which way to the lava?"

I wondered if the "experienced" driver/guide forgot the way each day. You'd think the way he drove, not watching the road, that he knew it well. I would keep tapping him on the shoulder so that he would miss large rocks in front of him. The woman in front also assisted - she would scream as the van went off the road! This was a signal for the driver to stop looking behind him.

Just before the end of the road, we had to switch drivers. Ours has not been deemed competent enough to get it up the slight hill. It was already dark when we reached the old lava beds. We followed a path through the dark, crusty remnants of previous eruptions.

After about fifteen minutes we could see lava glowing in the dark as it flowed down the volcano's cone. It gleamed red-orange against the nightfall. We climbed over hardened rivulets from previous flows, down crumbly valleys and back up the other side. After a further half-hour climbing, we could see eruptions spew out molten rock.

There was a half-moon overhead providing some illumination. We could hear the lava hissing as it cooled and little avalanches would rumble nearby. Sometimes fist-size rocks would tumble past. Some Germans refused to go further. Some Swedes pushed on despite the guide's outcry. They brought back still warm rocks. A faint, multi-coloured ring could be seen around the moon.

One of the students said, "It is the hole in the ozone layer". I thought he was making a joke, but the ensuing discussion revealed that he was serious. The others believed him. I explained that the phenomenon was the result of ice crystals at around 30,000 feet that reflected the moons rays much like a rainbow effect. The American was doubtful and insisted to his buddies that it was a hole in the ozone layer. I overheard him justify his diagnosis by questioning how could there be ice crystals in Guatemala.

Our return back through the lava beds was interrupted with "Oohs" and "Aahs" as each eruption outdid the previous one. It was well that we were fascinated with nature's pyrotechnic display as the guides did not know the way! We zigzagged up and down the slope sides for an inordinate length of time. There was more climbing before descending.

COPAN

To visit Copan in Honduras one of the (Learn) Spanish Schools organized a $40 trip. The bargain trip included six hours to get there, two hours at the site and five hours to return. This I learned from being on the tour. We had one flat tire and three breakdowns, including the engine quitting while crossing a flooded river. There were stops every eighty miles to put in three gallons of gas at a time.

The Maya site of Copan is known for its quality architecture and sculpture. It is not as big as Tikal, but it served an estimated 30,000 residents. The pyramids and temples are immense and elaborate. They are placed in a pleasing arrangement that makes walking about them extra enjoyable. One is able to appreciate the wide array of intricately carved stelae (stone monuments with hieroglyphic inscriptions), stairways, sculptures, wall decorations, and carvings. The outside is the results of the last rulers.

The Maya built on top of their predecessors. There are layers of pyramid and structures built over earlier buildings and foundations. There are some 700 meters of archaeological tunnels for exploring these structures but not open to visitors.

Our guide explained that it was because of fat ladies with high-heeled shoes who would damage the walkways. I noticed neither fat ladies nor high-heeled shoes among the tourists at the site that day.

Introduction to the Maya

The Maya were a complex society of MesoAmerica. I had taken courses about them as part of my Archaeology degree. Consequently I appreciated visiting the Maya sites of Cahal Pech, Xunantunich, Buenavista, Tikal, Caracol, Quirigua, Copan, El Pilar, Baking Pot, Altun Ha, Palenque, Chichen Itza and Uaxactun. Some my stories involve visiting those sites. Although not able to do justice to the Maya this is a brief overview of their complex realm and their significance to MesoAmerica.

MesoAmerica was an historical cultural area, from central Mexico to Costa Rica, which the Maya occupied a third of during their prime years. Their civilization existed in Belize, Guatemala, areas of southern Mexico and portions of Honduras & El Salvador. The Maya civilization prospered until AD 900. This date is based on the Thompson (GMT) correlation which may be out a number of years. They are known for their architecture, art, astronomical system, and calendar. Their civilization was sophisticated with a highly developed writing system. Individual city states were in contact with the Olmecs, Mixtecs, Aztecs and Teotihuacan at different times. Aspects of their culture are still present even after 500 years of European influence.

Maya sites

Of the hundreds of Maya sites ones with outstanding architecture include: Chichen Itza, Palenque, Uxmal and Yaxchilan in Mexico; Tikal and El Mirador in Guatemala; and Copán in Honduras. Although Caracol, in Belize, had an immense assortment of temples and pyramid structures, their architecture was not as significant or unique as other Maya cities. Difficult to reach, sites include Calakmul, Yaxchilan and El Mirador. Five day, mule packing, hiking trips are offered out of Carmelita, twenty miles north of Flores, Guatemala to reach El Mirador. In 1992, I was told donkeys would be needed to reach Caracol. It wasn't far from the truth, as noted in the story "Enroute to Caracol".

The Maya Civilization spread into the Belize some four thousand years ago. The middle and southern regions of Belize were dominated by Caracol, a large centre thought to have had 150,000 people. In northern Belize, the most important centre was Lamanai. Other sites in Belize include Xunantunich, El Pilar, Buena Vista & Cahal Pech (in the West); Santa Rita & Cerro Maya (in the north); and Nim Li Punit & Lubaantun (in the south). During the late Classic Era, it is estimated that one to two million people inhabited Belize.

Layout of Cities

Maya cities had a ceremonial and administrative centre surrounded by considerable residential sprawl. The principle architecture consisted of palaces, pyramid-temples, ceremonial ball courts and structures for astronomical observation. Plazas were used as meeting spaces and market places. Exteriors of buildings were painted and adorned with sculpture or painted reliefs. To archaeologists' viewpoints the cities went through slapdash expansion, with placement of additional palaces, temples and plazas in haphazard arrangements. Different parts were linked by causeways. In addition to growing outward temples increased as larger buildings were built atop previous structures. Some places had defensive walls.

Architecture

Among the vast array of structures would be an impressive acropolis. These were elite residential compounds with multi-room structures built upon platforms. Palaces were arranged around courtyards, with façades facing inwards. Some palaces had hieroglyphic descriptions identifying them as the royal residences. Courtly activities, such as receptions and rituals, took place in them. Temples were often placed atop a pyramid. Their buildings were distinguished by odd number of entrances: three, five, seven or nine, with center entrance being in the middle of a symmetric presentation.

Many cities boasted triadic pyramids that consisted of three superstructures situated on a plaza on top of a pyramid. The triadic pyramid was a common architectural form.   There would be a dominant central structure with stairways leading up from a plaza situated atop the pyramid, flanked by two smaller but similar edifices. Examples occur at Uaxactun, Caracol, Seibel, Nakum, Tikal and Palenque. The largest known triadic pyramid was built at El Mirador, which had three dozen triadic structures.

Stone stelae were dispersed throughout the cities. They were often paired with circular stones referred to as altars. At Yaxchilan, Dos Pilas and Copán sculptures adorned stone stairways. The hieroglyphic stairway at Copán, consisting of 2,200 glyphs, is the longest known Maya text.

Maya architecture incorporated various art forms and hieroglyphic texts. Examples were found on stelae (tall stone blocks) and ceramics. The Maya elite were literate, and had a complex system of hieroglyphic writing which they used to record their history in folding books called codices.

Codices

Codices were accordion like books that the Maya used to write down their observations, history, calculations and divinations. Only four codices are known to exist: the Dresden Codex (being the most elaborate, containing illustrated astronomical tables); the Madrid Codex (dated to between AD 1250 and AD 1450, the longest of the known codices, consisting of almanacs and horoscopes); the Paris Codex (containing prophecies); and the recently discovered Grolier Codex (of ten fragmented pages from a cave dated to around AD 1100).

Mayan

Before 2000 BC, the Maya spoke Proto-Mayan, a reconstructed vocabulary that likely originated in the Guatemalan Highlands. Proto-Mayan diverged into over 30 languages.

Mayan has a distinguishing hieroglyphic written form where symbols of pictorial meaning and sound syllables are combined in glyphs to represent words or phrases. Mayan text is generally arranged into double columns of glyph blocks. The glyphs are read in pairs in a zig-zag fashion.

Glyphs are composed of different elements that consist of a main sign and affixes. The major component may be a noun, a verb, an adverb, an adjective or a phonetic sign. Some signs are abstract, some are pictures of the object they represent and others are personification variants. Affixes are attached to a main sign and represented nouns, verbs or prepositions.

Mathematics

The Maya used a base 20 number system. This vigesimal system was also used by the Tsimshian on the North West Coast of North America. A dot would represent one, four dots four and a five was represented by a bar. Three bars with three dots above it represented the number 18. This bar-and-dot counting system was in use by 1000 BC. Keeping track of exchange of goods was important, where Cacao beans were used as currency.

Trade

Trade goods such as ceramics, textiles, copper, gold, silver, salt, stone axes and obsidian were carried by porters or moved in large canoes. Obsidian was lava that had cooled into glass like rock. It made for very sharp projectile points and elegant decorations.

Cities that controlled access to trade goods prospered. Trade routes also facilitated the movement of people and ideas. Shifts in trade routes occurred with the rise and fall of cities.

Politics and the People

Unlike the Aztecs and the Inca, the Maya did not form an empire. Calakmul, Tikal, Caracol and Mayapan achieved regional dominance in their times. The Maya area was a mix of states and chiefdoms that fluctuated in their relationships with each other. They were engaged in rivalries with interludes of servitude and coalitions.

Warfare in the Maya world was almost ongoing. Military campaigns were launched to control trade routes, to gain tribute and to take captives. They had classes of nobility, clergy, warriors, slaves and commoners.

Commoners, who were ninety percent of the population, built their houses from perishable materials that left little for archaeologists to identify. They engaged in producing subsistence crops and producing ceramics and stone tools; as well as making textiles and chocolate for the elite.

Dogs were domesticated by 2000 BC. Muscovy ducks and Ocellated (wild) turkeys were kept in pens. Rotational farming was used to grow crops. Terraced fields and use of fallowing were important for supporting outsized populations. The Maya diet was mostly maize, beans and squash, as revealed by pollen records, along with sunflower seeds, cotton, vanilla and cacao. Cotton was woven into textiles.

Time Periods

Development of the Maya is categorized by three major periods lasting centuries: the PreClassic (of about 2,000 years), the Classic (of 700 years) and the Post Classic (of 600 years). These in turn have been subdivided into phases like: Early PreClassic, Middle PreClassic, Late PreClassic and Terminal Classic. These periods were usually several centuries duration but could be as short as a century or as long as a millennium.

There was also an <u>Archaic Period</u>, prior to 2000 BC. This was the beginning of agriculture and the earliest villages. The Maya started as hunting/foraging bands that settled into small farming villages.

PreClassic Maya (circa 2000 BC – AD 300)

<u>The PreClassic Period</u> had the first complex societies with the cultivation of maize, beans, squashes and chili peppers. The first Maya cities developed around 750 BC. By 500 BC these cities possessed monumental architecture, including temples with stucco façades. Hieroglyphic writing was in use by the 200 BC.

In the Late PreClassic the city of Kaminaljuyu became prominent in the Guatemalan Highlands. In the lowlands the enormous city region of El Mirador grew to twenty square miles. El Mirador's largest temple covers six times the area of Tikal's colossal temple IV. Tikal was significant by 350 BC. Recent documentaries have stated that both these sites were twice as large as originally estimated. This is due in part to the use of Lidar to make 3D maps, a surveying method that measures distances using laser light in the infrared spectrum. This led to archaeologists being more receptive to how large ancient sites could have been.

Classic Period (c. AD 300–950)

The Classic Period, likened to the Renaissance, had multiple city-states, with populations around 50,000 to 100,000. These cities exhibited sculpted monuments. In this era city-states were linked by trade networks. In the fourth Century AD, the far away Valley of Mexico interfered at Tikal, deposing their ruler for a Teotihuacan-backed dynasty.

Calakmul, Tikal's great rival, was situated north of El Mirador. At various times one or other of these powers would gain a victory over its rival, resulting in periods of expansion for one and decline for the other. In the southeast realm, the important city of Copán was situated in Honduras. It also had connections with the Petén cities and Teotihuacan.

The AD 539 eruption of Ilopango Volcano, in Nicaragua, was one of the top ten Holocene eruptions on the planet. Worldwide the atmosphere would have been darkened for years. In MesoAmerica it would have caused crop failures and famine. The explosion would explain a demise of Maya activity around that time. People within 30 miles would have been forced to move as white ash buried landscapes in El Salvador, Nicaragua, Honduras and Guatemala.

Palenque and Yaxchilan were powerful in the Usumacinta region. In the highlands, Kaminaljuyu had become a sprawling city. Chichen Itza was the most important city in the northern Maya region of the Yucatan. It was in the Late Classic that Caracol established supremacy over Naranjo and won a war against Tikal. It replaced Tikal as the major regional power.

Collapse

The central Maya region no longer showed growth in the 9th century AD. There was widespread political collapse in the central Maya region. There was destructive warfare that led to abandonment of cities and ending of dynasties.

Reasons attributed for the decline have been: earthquakes, drought, environmental degradation, deforestation, warfare and combinations of these factors. Before the collapse there were hardly any trees in the area. Woodlands had been taken down to make lime. Much of the land was under agriculture or had buildings on it. Today, in Cayo District, you may note there are ancient stone ruins in backyards, visible on nature hikes and places with no roads along the Guatemala border.

During the Terminal Classic Period (AD 830-950), Chichen Itza & Uxmal, in the Yucatán continued to develop after the southern lowlands Maya ceased expansion. This apparent discrepancy may be attributed to the way the Maya calendar is interpreted.

PostClassic Period (950–1550 AD)
In the Early PostClassic Period the K'iche kingdom in the Guatemala Highlands expanded while Chichen Itza and its Puuc neighbours began to decline. The once-great city of Kaminaljuyu was abandoned after two millennia of continuous occupation. Cities became located on more-easily defended hilltops, with ditch-and-wall defences. New cities arose near the east and gulf coasts. Mayapan became a dominant power only to be abandoned in the mid-fifteenth century.

Astronomy
The Maya had recorded eclipse tables, calendars, and astronomical information that was more accurate than comparable knowledge in Europe. They observed and chronicled movements of the sun, moon, Mercury, Venus, Mars, Jupiter and the stars. They measured the Venus cycle with an error of just two hours. The codices noted that the eight cycles of Venus equalled five earth cycles (of the 360-day, eighteen month calendar).

Maya Long Count Calendar
The Maya calendrical system had its origins with the Olmec. The Maya developed it to maximum sophistication: recording lunar & solar cycles, eclipses and movements of planets with great accuracy.

The basic unit in the Maya calendar was a *k'in* (one day). Twenty *k'in* formed a *winal* (a score of days or a Maya month). The next unit, was multiplied by 18 to provide an estimate of the solar year (yielding 360 days). This 360-day year was called a *tun*. To make 365 days in a year, five days, called the *Wayeb*, were added to formulate the *Haab* cycle (a solar year).

The full Maya calendar counts nine cycles of time: *k'ins'* (days), *uinals* (months), *tuns* (years), *k'atuns* (twenty years), *baktuns, piktuns, kalabtuns, kinchltuns and alawtuns*. Each is twenty times longer than the previous, except for the *tuns*, which are eighteen times longer than a *uninal*. It is similar to the Gregorian calendar that counts five cycles: days, months, years, centuries and millennia.

Maya Months

In the Maya calendar there is both a *Haab* date (of the 365 day year) and a *Tzolk'in* date (of 260 days). The same date only appears every 52 years of the Calendar Round (unlike a Friday the 13th which can appear twice or thrice in our year). An event longer than 52 years required an additional cycle. This is where the Long Count Calendar was utilized. The Long Count Calendar gave each day a more unique designation, by counting five cycles of time (up to 1 *baktun*, or almost 400 years).

In the same way that our twelve months have names the eighteen Maya months have designations:

Pop, yellow	*Wo,* black
Sip, red	*Sotz,* bat
Sek, unknown	*Xul,* dog
Yaxk'in, blue (green)	*Mol,* day sign
Ch'en, black year	*Yax,* blue year
Sak', white (year)	*Keh,* red year
Mak, phonetic ma	*K'ank'in,* dog with torn ear
Mu(w)an*, hawk or owl	*Pax,* frog or iguana
K'ayab, macaw	*Kumk'u,* unknown
and *Wayeb,* sleeping period	

This is one Mayan labeling scheme. The English "translations" are my simplification from John Montgomery's 2002 <u>How to Read Maya Hieroglyphs</u>.

*the Muwan (or Muan or Moan) bird glyphs do have bird features and could be an owl or hawk. However I have argued that the so-called Moan bird, portrayed on a Lintel at Temple IV of Tikal (dated to AD 747), is not a bird but an octopus. They both have beaks. That is about the only similarity the image has to a bird. The lintel depiction shows eight arms of an octopus with suction cups. Not only is it an image of a creature from the ocean, it also displays the rules of grammar of CircumPacific Art, that is found in China, New Zealand, the Northwest Coast of North American and Peru (all in different time periods, centuries apart).

Maya Days

Like we have seven recurring days: Monday, Tuesday, Wednesday, Thursday, Friday, Saturday and Sunday) there are twenty Maya days that repeat:

Chuwen, monkey
B'en, unknown
Men, supernatural-being
Kab'an, earth
Kawak, rain (clouds)
Imix, water lily
Ak'b'al, serpent
Chikchan, snake
Manat, not known
Muluk, fish   and

*Eb,*wind
Ix, jaguar (Balam)
K'ib, univalve shell
Etx'nab', obsidian blade
*Ajaw,*sun god
Ik', wind
K'an, maize
*Kimi,*death *Manik',*
Lamat, Venus
Ok, dog

These twenty name days are portrayed on glyphs with a three-legged-stand holding up a cathode ray tube like icon. Inside are diverse images portrayed in various colours. Their name days were combined with 13 numbers, to make a cycle of 260 days called a *Tzolk'in* (incidentally equal to nine periods of the moon). Instead of having the first, second, third and fourth Monday of each month there will be 13 of each name day per *Tzolk'in*.

Long Count Dates

Examples of Long Count dates (based on the GMT correlation):

7.0.0.0.0	*10 Ajaw, 18 Sake*	(June 8, 354 BC)
8.0.0.0.0	*9 Ajaw, 3 Sip*	Sept. 7, 0041)
9.0.0.0.0	*8 Ajaw, 13 Keh*	(Sept. 8, 435)
10.0.0.0.0	*7 Ajaw, 18 Sip*	(March 9, 830)
10.10.10.10.10	*1 Ok, 13 Pax*	(Sept. 28, 1037)
11.0.0.0.0	*6 Ajaw, 8 Mak*	(June 8, 1224)
11.11.11.11.11 *2*	*Chuwen, 14 Sip*	(Sept 30, 1452)
11.13.11.14.0	*8 Ajaw, 8 K'ank'in*	(April 22, 1492)*
12.0.0.0.0	*5 Ajaw, 13 Sotz*	(Sept. 18, 1618)
12.12.12.12.12 *3*	*Eb,0 Ch'en*	(Oct. 15, 1867)
13.0.0.0.0	*4 Ajaw, 3 K'ank'in*	(Dec. 21, 2012)

* when Columbus landed

Notice all the long count dates ending with four zeros start on an *Ajaw* day (the Sun God). These long count dates covered nearly eight millennia (up to a *piktun*).

In recording dates it would be needless to include anything above a baktun (~400 years), as *piktuns* (~8,000 years) and higher would be redundant in their times. So they went to a Short Count calendar in the Classic Period, only extending a few centuries. Similarly we often only note the year of a century and not the century.

Therein lays the potential for confusion for us. Consider if someone found a document mentioning sailing across the Atlantic Ocean on April 19/19. Without further clues the finder could not be sure if the year was 2019, 1919, 1819, 1719, 1619 or 1519. Similarly the Maya took shortcuts in writing out dates and did not include all nine cycles of dating. There is the possibility that the commonly accepted GMT correlation is incorrect.

Correlation of the Long Count Calendar
The Maya started using an abbreviated Short Count Calendar during the Late Classic period. The Short Count is 13 *k'atuns* (a period of 13 x 20 = 260 years). Consequently there is potential for ambiguity on how the Maya calendar corresponds with our (Gregorian) calendar. Over fifty correlated are postulated, some vary by a day or two. The most generally accepted correlation is the Goodman-Martinez-Thompson (GMT) correlation.
The Spinden Correlation has a Long Count of 260 years behind the GMT correlation; it matches documented evidence as well as the archaeology of the Yucatán Peninsula. The George Vaillant Correlation would shift Maya dates 260 years ahead, resulting in a shorter Post Classic period. Some radiocarbon dating supports the GMT correlation; some support the Spinden correlation.

Antoon Vollemaere argues for a 520 year difference (2 x 260) to the GMT correlation. He provides examples of finds (e.g. interacting with the Aztecs, styling of vases, manufacture of copper bells) that would fit this dating better and that all the 69 mentions for solar eclipses are accounted for (whereas only half align with the GMT correlation). Thus Vollemaere suggests there was no large gap in the Maya empire – only demise related to the onslaught of Europeans.

The GMT correlation yields the famous Maya December 2012 apocalypse date. Presumption about end of the world arose from translating an inscription, believed to read: "On this date a great event will happen" that corresponded with a Maya Calendar cycle ending. Disaster didn't occur (except in the movie). Using the Spinden correlation Armageddon was slated for 1752. That was the year Britain abandoned the Julian calendar for the Gregorian, not too much of a catastrophe. The Vaillant correlation would put the year of reckoning at 2272. That year is still to come. With the way we ignore climate change and environmental warnings humans may not be around then.

Using the Vollemaere correlation of 520 years from the GMT correlation yields the date 1492. That was inauspicious for natives of North and South America. Hundreds of millions perished in those continents after Columbus set foot in the New World. Cataclysm indeed.

1993 – The Early Classic Period

Do Not Stand Under the Monkeys

January 1993: Manfred and his family bought the mansion I had been staying in. We switched residences. I moved into Manfred's old place.

THE BLUE HOLE

Because I took the two Lohr brothers canoeing for "their" birthday, I took two of their sisters, Naomi, and Lucia to the Blue Hole for their birthday. There are two Blue Holes in Belize. One that is quite distant from shore, as part of the barrier reef - a now submerged limestone cavern, popular with experienced scuba divers. The other is inland, along the Hummingbird Highway.

Dr. Pat took all of us in her van. We stopped first at St. Herman's cave, one mile from the Blue Hole and examined some caverns - not as spectacular as Rio Frio Cave, but then we did not raft ride through the tunnel.

This Blue Hole is a small crystal-blue swimming pool. Only five at a time are allowed to swim. The water enters the pool from a magical looking waterfall twelve feet up and leaves through a stream that disappears into a tunnel leading to St. Herman's cave. From inside the tunnel, at water level, it looked like a scene from some tropical paradise.

Postscripts 2019: *Lucia is living in Arkansas; Naomi still lives in San Ignacio*

THEY CALL MY CHICKEN "HE"

Manfred had left twenty-three chickens at the farm; they were hardly laying at all. So Manfred bagged them up to take to the market where they would fetch $3 each. I arrived home just as he was carrying the bags to the vehicle. I pleaded with him to let me have one for $3. He relented and picked out one that he thought would be a good "layer".

For some reason, they refer to my chicken as a "he" or "him". Maybe there is something to referring to chickens as "he's" but I would have thought if you are not going to call it an "it," then it would be a "she". As it turns out, "he" has yielded five eggs in six days. "He" lays three days in a row then skips one day - about every 32 hours - more than enough eggs for me. Seven more eggs and I break even on my "investment".

NON–VERBAL COMMUNICATION

Every once in a while I experience some gratifying instance of communication where no words are exchanged. One instance was while walking home, after dark. Manfred's cows were still at his old place. I had wanted to be back before Manfred finished milking. I pay him in advance for ten quarts. It was time to pay him for the milk he leaves me each day. That Monday morning I had left a two-quart jar by the milking shed.

As I headed down the long hill, I could hear the "putt-putt" of his Volkswagen coming up the hill. Manfred counts on me paying in advance, and I knew he would be looking for his money. Even though I could not see the trailer, he was pulling I knew he would not want to stop on the hill, as his VW needs a good run to make it up any hill.

Manfred knew I knew this. As I was walking down the middle of the road with my left hand held out to the side, he knew it was not a wave. In the fraction of a second between regaining my vision due to his headlights shining at me, as Manfred passed, I saw his hand level with mine. The ten dollars passed to him. With the hand-off complete he hollered "thanks".

Okay, so it was not completely non-verbal, but it reminded me of a totally non-verbal communication that took place at the Calgary Stampede in 1964. I was throwing toilet paper rolls through the toilet seat hanging from the booth's ceiling. At twelve years old, I was not good at throwing footballs through tires, nor getting basketballs through hoops but man I could get that roll of toilet paper through the toilet seat almost every time.

There I was, collecting little stuffed toys, worth two bits, for every three rolls I tossed through the toilet seat. Throws were 3 for a quarter. For every three little stuffed toys, I could trade them for one a little bit bigger (worth half a dollar). I could trade three little bit bigger stuffed toys for a medium size stuffed animal (worth 75 cents) and so on until about the tenth trade up were five big stuffed animals garnered one bigger stuffed one.

I was not in for trading so much as winning (I calculated that I was better off with quantity and not quality). There I was, getting a stack of these stuffed toys (which were looking uglier and uglier) as the guy behind the counter dug them out of some hidden box. A bit of a crowd had gathered to marvel at my performance. There was a girl, about eight years old, to my left. She looked at me, and without saying anything, pushed a quarter towards me.

Without saying anything, I pushed the quarter to the man behind the counter. Without saying anything, the man behind the counter took the quarter and pushed three toilet paper rolls at me. The first went cleanly through. The second went in as it brushed the left inside edge. My arm went back for the third throw; the crowd was hushed (have you heard a hushed crowd on the midway?). I looked to my left; the girl was tightly gripping the counter, which she could barely see over. I threw.

This was no "Casey at the Bat" story; the toilet roll went through the toilet seat. The man gave me another ugly stuffed toy. I handed it to the girl. She hugged it to her, looked momentarily at me with a sparkle in her eyes, and then skipped into the crowd before I might change my mind. I did not throw any more toilet paper rolls. No quantity of stuffed, ugly toys were going to be more rewarding than the joy on that girl's face.

EGG PICK UP

February 1993

On February 9, 1993, I drove into Belize City (two hours travelling time) to spend a day getting the paperwork done to pick up hatching eggs from the airport. The eggs were arriving February 10th, for the new Crystal Hatcher's project. I started at the Agricultural Department, to get the quarantine stamp, but they told me I had to go to the new customs building out by the port. When I reached there, the Supply Control person told me I had to go to the Chamber of Commerce downtown first, but I could not go until the afternoon.

It was closed. I waited and waited. Some guys came along and informed me that the office had been moved to Belmopan (the capital, halfway back to San Ignacio). So I went back to the Customs building. They apologized for sending me to the closed place and put a whole mess of stamps and scribbles over the umpteen documents necessary to get the eggs released.

I specifically inquired if this was all that was necessary to which I received an affirmative reply. I asked again in a different way "Does this mean I can pick up the eggs at the airport now?" They said "Yes" but then after a moment "...but you might check with the Quarantine people."

Sure enough, the Agricultural Office had to add his stamps, scribbles and fill out forms. He too said that now I could proceed to the airport. It sounded good at this point.

I wanted to get those eggs as soon as they were off the plane and to the incubator. I even had time to check out the Craft place for placing the leather and sewing groups' products.

Next morning I set off to the airport (a half hour drive north of Belize City) and proceeded to the Quarantine Inspection - no problems - more stamps and forms filled out, but I was given a go-ahead to clear with Customs.

It was a dead end. The Customs bloke said the forms were not filled out properly; I informed him that I had spent the better part of the day getting stamps and scribbles and numbers and notations - to no avail. Whatever forms, stamps, notations, scribbles, numbers were there, were not what he wanted to see.

He insisted I had to go back to the Customs building, by the port, a half-hour there, a half-hour in line and a half-hour back. It was not the half-hour to get there, nor the half-hour waiting, nor the half-hour coming back that got me, nor that were the eggs probably hard boiling out on the tarmac.

It was that the "new" CCDC's truck's gas gauge was not working. I would have had enough gas to get back if I had not had to do all the extra running around. I had come in a day early to get the paperwork out of the way and asked the right questions, but I was not accomplishing my task. I was frustrated. When I finally picked up the eggs, all the guard wanted to see was my driver's license (none of the paperwork).

THE ART OF SMUGGLING

February 1993

One Thursday, I went to Chetumal, Mexico, for cloth for the sewing group to make T-shirts. The previous week I had been to Melchor, Guatemala and purchased all the shirt able-cloth there. The Field Officer, President of the group, and me had divided up the thirty yards of cloth and straggled through Customs separately in a ploy to avoid paying the 30% duty.

Sacred Heart College wanted yellow shirts for their sports day competition and offered the contract to SIWA. As there was no more suitable cloth in Melchor (yellow or otherwise) and none in Belize City I headed to Mexico. It was a four-hour bus ride from Belize City to Melchor. At the market, I took a taxi to the San Francisco Shopping Centre to a sewing specialty shop.

I was wandering among dozens of Mexican women, trying to explain, with my eight days of Spanish training, that I needed *"mucho"* material for T-shirts. It was a fine time to realize I did not know T-shirt material from cotton for jeans. The sales representative directed me to a section where I bought all the yellow cloth in stock. The next feat: to get the cloth across the border.

Sitting at the back of the bus were three women, each with many bags full of items. I had watched them "fumble" through customs (as I was held up with my "purchase") moving bags slowly with a show of awkwardness.

After customs, back on the bus, the women took out the flattened boxes they had stashed under the seats and began to replace the goods in their wholesale cartons. Among the three of them they had managed to get through two cases of cooking oil, an equivalent amount of hair shampoo, detergent, marshmallows, assorted candies, and other goods.

They had bought wholesale, mixed the stuff on the bus in different size and coloured bags to look as if they'd been shopping. It was amazing how many boxes they filled up. They had their smuggling down to an art. My technique was a little different. I had a letter from the school principal pleading to allow the cloth to be brought in for the school, implying the T-shirts would not be resold. This was technically correct and allowed the sewing group to fill their contract.

CHICHEN ITZA AND THE JAPANESE LOVE LETTER

March 1993

In March I took some vacation time to travel to Calgary. I needed to make arrangements in ordering a prototype of my patented computer keyboard. Thinking it would be cheaper, I flew out of Cancun. Later I learned that I could have flown for the same price out of Belize City had I used a Belize City travel agent. I had an extra day in Cancun and used it to visit Chichen Itza, a spectacular Maya site about 200 kilometers away across the Yucatan peninsula. Tour buses went there daily at a cost of $105.

It was too steep for my budget. I thought I would be clever and take the regular bus ($5 US). I thought I was going to come out ahead by paying $5 there and $5 back. The bus departed at 7:30 a.m. and dropped me off at 12:30. It was a long slow ride. Five hours going along a bland, flat countryside - maybe the higher tariff would have been worth it.

QUITE THE SITE

Chichen Itza was certainly worth seeing. On a scale of 1 to 10, I would give it an 11 (extra marks for the cenote). There were numerous tour buses discharging visitors to compete with, in moving through the turnstiles. It was two days after the spring equinox, and I hoped I might catch a slight version of the "shadow serpent" moving along the main pyramid steps. Twice a year, during the equinoxes, the placement of the pyramid and design of the steps "causes" the shadow of the steps to resemble an enormous serpent traversing the stairs.

The day was hot with the sun beating down. Fortunately, a soft drink/souvenir stand was strategically placed, as a welcome sight. At one stand was one of the people who had taken a tour bus from Cancun. It had also left at 7:30 a.m. and arrived when we had. Their bus had stopped at predetermined souvenir stands.

By keeping to the right, I thought I should be able to take in the entire site in an orderly fashion. That started me with the "astronomy observatory" and other structures in that sector. The observatory is round and offered a variety of entrances to explore. A little way off was a more intriguing structure with numerous archaeological tunnels to explore.

On the outside of one structure, I was entranced by the interacting motif of symbolic faces molded into the corners. Viewed from one side, it made a face with the "nose" hanging out of the corner. Viewed from the other side the same nose was part of a different profile - much like Northwest Coast boxes. These noses also served as large hooks that could have been used to hang flower pots.

One of the main attractions at Chichen Itza is the Kulcan Temple, an enormous square pyramid in the center of the complex. There are huge double stairs going up all four sides. People crept up slowly. I took the stairs two at a time and was soon at the top - breathless (and not because of the view). I enjoyed walking around the top, noting the interesting carvings and use of colour. Getting down was another matter. I noticed others having trouble going down. When I looked down the stairs, I realized why. What a sense of vertigo!

There was no way I could go down and look down at the same time. How did anyone get down? I watched several try. They went very slowly. They moved from step to step by sitting down and then sliding to the next stone.

I could not spend all day there, and I was sure no helicopter was coming, no matter how long I doddled. I went down the steps at an angle, slowly, slowly, slowly, moving one foot to the next, never looking further than two steps away. Not like the way I went up! Back on the ground I looked up and shook my head. I watched a group of tourists argue about starting up and wondered if they knew that coming down was more of a challenge.

CHACMOOL FOOL

Next, I studied the stone Chacmool figures, statues about twice human size, of a rain god on its back, half sitting with knees bent, holding a bowl on its stomach. It is the motif used by the Archaeology Association at the University of Calgary. It also looked very much like a potlatch bowl used by natives of Vancouver Island, the same size, posture and tilt of the head. There is no similarity one of my Professors, at U. of C. assured me. I was just being fooled.

More intriguing were the hundreds of columns set up like a forest, row upon row - similar to the ones at Karnack in Egypt, but not as tall. This was a good place to explore. On the other side of the plaza was the largest ball court known in Meso-America. The information plaque said it was covered when in use (two football fields big!).

It had clouded over, making the air a little cooler. Unfortunately, that meant no shadows and hence no chances to make out the serpent snaking along the stairs. I paused to explore the museum and gift shop (without making a purchase). There was a scale model of the site. In addition to increasing the appreciation for what was visible now, it showed how it might have looked centuries ago. It also revealed the location of a large *cenote* (water hole) near the center. There were no signs to find the cenote via the trails.

CENOTE

Following a slightly used path, I headed in the direction of where the cenote would be located. I was rewarded with a view of the edge of the cenote, a large, submerged, circular, lake-like depression about 200 feet in diameter. The top of the water surface was about 80 feet below.

I was able to walk all the way around (circumventing fences) and found a way down to the surface. There were a number of caves along the way that went back 20 - 40 yards. Studies done by archaeologists investigated the myth that it was used to sacrifice maidens. Skeletal remains discovered there were deemed consistent with quantities from accidentally falling in.

BLIND SIDED

The bus ride back was noteworthy in that it only took three hours. I wanted to tip the driver but was not sure that a gringo should reward fast (equate that with hazardous) driving. As I pondered the morality of wanting to show my appreciation a solution presented itself. The driver had allowed a blind man to board who sang for us. As he worked his way down the aisle, passengers slipped him money. I gave him the 5,000 peso note I had thought of passing to the driver. It was the driver who let him on so he must want to help out. There would be no way for him to know which passenger gave how much.

SOMETHING TO REMEMBER CANCUN BY

Returning back to Cancun in the early evening I had time to visit souvenir shops, perfecting my "no thank you" to various pitches to buy: carvings, hammocks, drugs, jewellery, and other assorted items. As I went by one shop, the owner-operator had a curious pleading to his voice, so I accepted his offer to browse. We talked (using his not so bad English and my poor Spanish) about his business and about learning different languages. As I sat in his shop I experienced the perspective of shopkeepers watching tourists rush by; I understood why Jose wanted someone just to stop and look.

With thousands of shops, it was hard to get market share located at the back. It turned out he had other cares than someone to just look at his goods. He had received a love letter from a Japanese woman he had met. The letter was written in English. He needed help reading it, and more importantly, help in composing a reply. I was touched that he would be willing to let me read the letter but a little confounded about helping him write a response.

Jose told me that I was the only one, in two weeks since the letter came, who had bothered to stop that he could ask for help. His Mexican compadres were "not-so-good with the English".

He had met the Japanese woman while she was on vacation, in Mexico. They had fallen in love. On the day she was leaving he was supposed to call her, but he left it too late. She cried all the way to the airport when she did not get his call. She could not leave it at that, so she wrote, offering to come back to live in Mexico.

He worried about whether to call her or not and did it too late. He wanted to express in his letter that he would be overjoyed to have her come back to Mexico and they would get married. He had drafted a one-page reply that had taken him four days to write. I told him that it was good the way it was and that he should send it.

But he wanted it to be grammatically perfect. I tried to convince him that this was not important. I relented to his persistent persuasion and "fixed" the letter, embellishing it as I went along. After he signed the bottom, I realized he intended on sending my version. I rewrote it again neatly (no small feat for me) enhancing it more after Jose showed me photos of her, her family and their home. On one hand, I was concerned with the authenticity or ethics of it. On the other hand, the fellow's desire to making the most favourable impression possible assured me there was no way I could overdo it. Not that she needed much encouragement, but this may have been a very significant letter.

Postscript: _December 1993_: _Again in Cancun, I checked up on Jose. He sadly informed me that his wife had come into the shop, took all the letters, and sent the Japanese woman a photo of Jose with his five kids._

ATTACK OF THE COCONUTS

June 1993

June must be the month coconuts take leave of their trees. Throughout the night I hear them hurtling earthwards succumbing to gravitational pull or yearning to take root. Sometimes they land on the metal roofs, giving the impression of a cannonball launched from the next valley. Other times, they will be sitting along the pathways with a crooked smile as if to say "get you next time".

The coconuts are not the only projectiles of the night. The most disconcerting are the streaking fireflies. I can see them from some thirty yards away heading straight for my head, leaving a lightning trail like some tracer bullet in exile.

Frogs dominate at night. The bullfrogs give no quarter as I approach; the sticky frogs launch any which way. Being sticky on all sides they merely have to make contact to stick. One frog, in the main water tank, makes an echoing croaking.

The other frogs sing until they hear this frog harping in. The peculiar booming makes the others stop. I have endeavoured to remove it but cannot locate it. Either frogs migrate, or they have been scared off by the lone croaker. If it is amphibians by night, reptiles rule by day. Lizards abound, and they truly do leap Batman! I am impressed every time they make an agile leap or scurry through the bush.

One morning a snake put in an appearance around the shower. Bats flap at night and more birds than I can name fill the thickets making various calls from humming to monkey-like squawking. I recognize hummingbirds, owls, herons, cicadas, melodious blackbirds, parrots, and toucans.

The jungle growth slowly makes incursions into the pathways connecting the buildings. Vines fill most spaces in the kitchen area. I wrapped more and more up in the rafters and wondered if their weight would bring everything down. One time I had cut a vine and the eaves trough crashed down. Somehow it was intrinsically bound with keeping the piping in place. It took quite a while to get it set up again. I had the solar panel set up with the wiring hidden among the vines.

THE UNIVERSAL FOOD

The farm was invaded with tens of thousands of flying termites. They seem to seek heat. Apparently, they come once a year, at the end of the dry season, flying aimlessly about until they land and shake their wings loose. Towards evening the birds were snapping them up in the air until the bats took over. Along the walls, lizards munched on plenty, and near ground level, frogs feasted. Termites making surface contact became fodder for all sizes of ants. Normally, I set the scorpions and spiders away from the house and cabin, but there were too many gobbling up termites. Besides, I figured they would be well fed and not bite me. Even fish in the ponds were filling up. Insects are a universal food.

DO NOT STAND UNDER THE MONKEYS

June 1993

I had not been on a horse for five years, and an all-day ride was not the way to get re-acquainted. Charlie has lived in Belize for more than a decade. She runs a budget riding service called "Easy Rider". Jan, a horse enthusiast from Oregon, had booked a day ride and Charlie likes to have at least three riders per trip, so she invited me along.

The rainy season had started, so I was warned to bring my poncho. It had rained solidly the previous two days prior to departure. Fortunately, we had sunshine to start the trip. We set off, Charlie in front, Jan in the middle and myself in the rear. Charlie had not been out on the trails for three weeks so a lot of growth and changes could be expected.

We were surrounded by forest for miles after leaving the road in front of Easy Rider. Charlie swung her machete to clear new growth and fallen branches. Loose vines and branches hung down which I broke off when they were within reach. Charlie "requested" I refrain from using my machete as the horses' ears might be severed as easily as everything else sticking out. We rode well-groomed trails kept by Windy Hill Cottages.

We came across a number of leaf cutter ants carrying bits of leaves; Charlie explained that they did not cut the leaves from around their "city", as they need shade to keep the temperature constant for their fungi farm. She also pointed out their architecture.

When it rains, the water does not flood out their homes or fungi growth areas. In other places, Charlie pointed out various jungle trees and explained how they could be used for ailments. One interesting pair was the gumbo-limbo tree and the black poisonwood. They are usually found together. The healing part of the gumbo-limbo comes from the black poisonwood which gets its poison from the gumbo-limbo.

In some places, the paths were wide enough and long enough to gallop the horses. In the morning the horses are eager to run. We came out of the jungle on a hillside overlooking the Mopan River valley. Formerly the tallest structure (displaced by the discovery of Caana pyramid at Caracol) in Belize, the Castile, at Xunantunich ruins, could be seen in the distance along with numerous small farmhouses.

We proceeded through orange orchards and then back into the jungle. We dismounted where Charlie usually finds a troop of Howler Monkeys. After walking in circles for a while, Charlie summoned me down the trail. Sure enough, there was a Howler Monkey picking fruit near the tops of some trees. The sun was behind it; Charlie had named it Philippa.

After the monkey moved, we realized it was a Philip. It was quite graceful and seemed to be unconcerned with our observations. We followed as it led us to five others.

One was a baby. Charlie stepped under a tree, to get a better view. "She's peeing on you," I warned. Charlie stepped back and remarked, "I've been told they will throw shit."

On cue the monkey defecated - communication from one species to another letting us know we were not welcome!

We left and rode to the river to have lunch. We rode the horses into the river so they could drink. Surprisingly, my horse, Coco would not drink. Normally this would also include a swimming break, but due to the recent rains, the river was too high and moving too fast. So we just enjoyed a leisurely lunch.

The next stop, further upstream, was the Maya site of Buena Vista. It is one of the bigger sites in Belize - 45,000 occupants in its time. Caracol is thought to have had 190,000 residents. Buena Vista is vast. The site is on grazing land, which presents a better sense of how spread-out the various components were. We could discern temples, pyramids and mound structures. On the long trek back we spotted pottery shards where the earth has been churned up by horses or vehicles passing over the area. There were open spaces to run the horses, but Coco was not too keen on keeping up. That was okay with me as I had a sore posterior and aching knees. I was content to mosey.

Postscript 2019: Charlie now has a craft shop on Burns Avenue and is President of the local Rotary Club.

TO CATCH A CAT

One of the Lohrs brought me a stray kitten. At first, I was unable to get near it, but once it figured out who was providing the twice-daily ration of food, it would not leave me alone. I named it Sylvester. I tried to explain to the Lohr children about Sylvester the cat, on Bugs Bunny. The only Sylvester they knew about was Stallone, and they wanted to know why I named my cat after Rambo (besides it's being rambunctious). It turned out to be female, so I renamed her Sylvie. Despite my admonishing Sylvie about being allergic to cats she crawled onto my bed at night and would slowly sneak up to my chin.

Catsup, one of the cats the Lohrs' had, before moving, showed up. It meowed most of the night. Sylvie only took a minor interest, but the racket drove me crazy. I set a trap with a light plastic box. Propping one end up with a screwdriver, it was designed to enclose a cat eating food inside. Several nights went by with no Catsup in the "trap", but the food was gone.

One day, while I was in the kitchen, Sylvie got caught. At least now I would be able to see if it could hold a cat. Sylvie tried to get out by putting her paws through the holes in the case sides. Then she positioned her head against one end and pushed. She got that box moving around so that it appeared to be a breadbox under its own locomotion. She headed for the stairs; that released her, but the box still had momentum. It looked like the box was chasing Sylvie down the stairs. That was pretty funny! The box was not going to keep Catsup locked up.

Catsup no longer meowed at night and was brave enough to dash in while I fed Sylvie. One time Catsup gave Sylvie a smack in the head while "sharing" the food. That was the last time Catsup got the better of Sylvie.

Although only half her size, Sylvie pounced on and otherwise tormented the bigger cat. Sylvie even began to place her left paw across the feeding dishes so that Catsup could not feed at the same time. Even when I set out separate dishes, they both ate from the same one to make sure neither received something the other did not. They also put frogs in their mouths. I tried to discourage that activity.

Catsup became a member of the household. Now both of them follow me everywhere I go. They follow me to the kitchen; they follow me back to my cabin, they follow if I go to the main house. When I go to the outhouse - it's a threesome.

My hammock hangs outside the cottage under the porch roof, with an abundance of plants, flowers, and trees growing around it. I watch the birds fly around during the day and view the stars at night. One day while reading in my hammock, I heard a squawking on the porch. I saw what appeared to be a lizard with a very long body. I squinted to determine what this noisy beast was. It was a frog with a snake attached to its rear end!

Somehow the snake drew itself backward, through a small hole. Just as the head of the snake (and head of the frog) was about to disappear, Catsup, having noticed my attention focused on the retreat, jumped up and smacked that snake in the head knocking the frog loose.

You might think that the cat was saving the frog by scaring the snake away. Nope! Catsup popped that frog in her mouth. Darn cats! - cannot stand to see something else have something in its mouth. I chased the cat. The snake got away; the frog got away; the cat got away.

MANFRED'S DOG BIT ME

June 1993

Charlie was in Sweden to help her daughter settle into college. I was asked to look after the place for a week. Charlie lived west of town on the Bullet Tree Falls Road; I lived south of town, on the Cristo Rey Road. One of the Lohr's had borrowed my mountain bike. I needed it to go back each day to feed Catsup and Sylvie.

When I arrived at Manfred's place, I greeted the dogs: Woozo, Garth, and Tobie. They stopped barking and wagged their tails when I patted them on their heads. After all, I helped raise Toby's puppies. I got the bike out of the shed and as I was walking to the gate "CHOMP", Woozo bit HARD into my right calf. Good thing it was one of the times I wore long pants. I could not walk properly; I could barely hobble because Manfred's dog bit me.

By sitting on the bike and pushing with my one good leg, I got to the CCDC office where I borrowed the motorbike to reach Charlie's. I could not get around to checking on the horses, because Manfred's dog bit me. The next morning I fed the dogs.

Chaka, a cross between a German shepherd and a grizzly bear, is the alpha dog. The other dogs walked all around me, stepping on my heels, stopping in front of me, bumping me on the sides and I could not get out of the way ... because Manfred's dog bit me.

That night I tried to feed the parrots. The dogs just hovered around, and one parrot squawked up such a storm it sent the dogs into a tizzy. I could not chase them away ... because Manfred's dog bit me. I just stayed inside and read. The dogs scratched at the door and seemed to bark all night. I got up five times to see what the commotion was but did not see anything in the dark.

Sometimes they were just chasing Limpey, the horse with the sprained leg, kept in the inner yard. I could not go far beyond the main yard where the dogs were locked in ... because Manfred's dog bit me. The next morning I had to deal with the horses. Normally they would be taken care of by Manual, the hired hand. But he had left his wife and ran off to Guatemala with a 13-year-old girl.

Other than Limpey, there were twelve horses; all were waiting to be fed. I had been told that each horse would feed at its own station. I let the horses in and tried to tie them up at their feeding spots. That worked for one horse, but the other horses would not let me put ropes and halters on them. Some would not even let me get near. I figured, okay, I'd put the grain mix out in their respective feed boxes.

That worked for two horses before the other horses started fighting over the feed. Now I could see why the feeders were positioned where they were. I was not going to be outsmarted by horses! By shaking some feed into a bucket, I led the horses out into the pasture, closed the gate and led them back one at a time. Once the feed was in their feeder, I could put a rope around their neck.

Then I got another one, which was tedious business with me hobbling (because Manfred's dog bit me). The horses nudged me, pushed me and even bit my hand but (damn fortunately) none of them stepped on my toes (not like the dogs). I also knew not to get behind them to avoid any well (or poorly) placed kicks. One horse, the ringleader, I could not reach. As long as it had grain in its feeder, it did not rob the others.

After they were all fed, I had to move them back to the pasture. I took them one at a time and tried to make the gate work like a one-way valve: horses into the pasture, none into the backyard. However, they kept congregating around the gate, and I could not maneuver very well (because Manfred's dog bit me). I moved them all except the ringleader. It would not let me near.

After a hobbling pursuit, I had it cornered with a rope around its neck. Have you ever been in a tug-of-war with one thousand pounds of horse? I was not the winner. By wrapping the rope around a post, I had leverage to stay even. Force was not going to work. I opted for gentle persuasion - talking softly and moving slowly; it followed me out.

While walking past the garage, I noticed the storage shed was open. Someone had broken in and rummaged around. As not too many locals knew the dogs did not have access to the back yard I figured the hired hand that had returned - he knew what was kept in there.

Because Manfred's dog had bitten me, I could not walk into town to report it to the police. Like my place, Charlie's did not have a phone or electricity. I tried to start the truck. The battery was good, but it would not catch after many attempts.

I leaned over the steering wheel in exasperation. From this vantage point, I saw the choke (pulling that made all the difference). While in town I tried to "round up" someone to cut the grass. Money had been left money for that. It would take three days to mow all the lawn. I could not do it ... because

Manfred's dog bit me. I also fed the cats and Baby. Baby is a heifer, ready to breed. I would fill her bucket with feed and position the bucket over the fence. Baby moved the bucket away from the fence so that in order to feed her I had to climb over the barbwire fence to retrieve the bucket.

At the sight of me approaching, Baby came a running. For some foolish reason (from being around Manfred's skittish cows) I figured she would stop before getting to me.

Have you ever wondered what it would be like to grab a steer by the horns? Oomph, the beast butted me square in the stomach. I could not get out of the way - because Manfred's dog bit me. The next day I worked the feeding of Baby with two buckets.

At 5:00 a.m. the next morning the dogs woke me with some fierce barking. Getting up to investigate I discovered the backyard full of horses. They had broken down the gate.

I figured that would take care of mowing the lawn and went back to bed. Besides I was not going to be able to return them to the pasture - because Manfred's dog bit me.

Chaka could be really mean at times. He picked on the other dogs, especially the pups, even though they were the size of full-grown Labradors. Chaka also terrorized Limpey. One night after midnight, Chaka was hounding Limpey around the inner yard. Limping myself, I tried to catch Chaka. I was carrying a big stick but could not get close enough to interfere.

When the horse ran by in the dark and the rain, I made a judgment of the speed and direction from Chaka's barking and snarling of where to throw the stick so it would meet up with where the dog would be. After making my calculation, I swung and let go. The barking stopped. Limpey, my fellow cripple, was no longer being chased. From all the hobbling around my leg was painfully sore, where ... Manfred's dog bit me!

VALLEY OF THE PARROTS

June 1993

Now that the rainy season has started, several hundred parrots have gathered in the hilltops of the valley where the farm is situated. They have been flocking in by the dozens, and the racket they make transcends the muffler-less vehicles climbing up Devil's Hill. Every year they come to this valley. When Manfred lived here, he thought they came to all the valleys. Since he moved, he has learned that this is apparently the only one in the vicinity. The parrots gather to find a mate.

During the day bunches of birds fly back and forth across the valley, from tree to tree. Toward sundown, hundreds can be seen taking off, in groups over the farm. The further they get from take-off, the more dispersed they become. As they fly overhead you can see that most of them are in two's, probably to build a nest and start a family. Shortly after the first batch flies over, another wave of a hundred or more cross-over. This happens several times before dusk.

We only see the parrots flying over us; it may be that several hundred more fly in other directions. This procedure repeats, over and over for several days. It is not known (by Manfred nor me) if it is the same birds trying again each day, or if different ones arrive on other days during the mating season.

THE BONES OF DON FABIO

July 1993

A sergeant from the Benque Police came to the CCDC office to report that human remains, about four weeks decomposed, had been discovered on the CCDC Ranch. Juan Sanchez, the Field Officer, had been there two weeks earlier, to pay Don Fabio the property caretaker. The officer wanted him to come out and have a look. I did not think it was a good idea for him to go alone and said so to the CCDC Coordinator. The next day the three of us went with the sergeant. I went along more to provide counsel, should the officer keep up the insinuations about him being the last person to see Fabio alive. If indeed it was Fabio!

We found the remains of a skeleton scattered around 400 square feet of a small clearing. This was in the lower part of the property, where there was no reason to go. Juan recognized Fabio's cap, boots, and pants. The pockets were turned inside out. The skull had no teeth. Fabio had no teeth.

The sergeant was going to throw the sternum away, as it did not look like a bone. I explained to him, while pointing to his chest that it was where the ribs connected. Who would have thought that sitting through a human osteology course would mean offering this analysis? Ribs were picked out from wood scraps of the same colour; toe bones were distinguished from finger bones. The vertebrae column had a pronounced bend. Fabio had walked stooped over. The one femur we found measured 17" which corresponded to a man of 5' 5".

That was Fabio's height if he could have stood up straight. I recommended collecting the remnants of clothing as that would be the only items showing signs of blood.

We went to the buildings at the top of the hill, on the edge of an ancient Maya temple, to discover his rifle and machete missing. He always kept them with him, and they were not found near the "body".

The last time I saw Don Fabio he was smiling a toothless grin, cooking lunch for the harvesting crew and visitors. His beans and rice were the best I ever tasted. The sergeant talked about what to do with the dozen or so chickens. Every once in a while the sergeant asked an oblique question, relating to the death, to see how Juan would respond. He found it hard to believe that he had seen Fabio alive just two weeks ago.

Juan must have found it hard to accept Fabio's death. I had trouble dealing with it. Fabio was a smiling old man that was living out of the rainforest just two weeks ago, and now there were just bones. The jungle takes care of its own.

Postscripts: *A week later a detective came by the office to say bullet holes were detected in the shirt, found with the bones. A forensic specialist had been called in to examine the remains.*

Two days after that a person approached Juan at the Benque Fest telling him Fabio was killed over a dispute about timber rights. A week later, a rumour was circulating that he had been shot trying to prevent looters from removing artifacts from the Maya temple.

BACK AT THE CCDC RANCH

July 1993

After the killing of Don Fabio, the CCDC staff mounted an expedition to procure the chickens. They could not stand the idea of a dozen chickens running around free with no one to "eat" them. The Field Officers borrowed rifles and set off early with the Board Chairman.

At 1:00 p.m. we received a message at the CCDC office. The CCDC truck was broken down; the battery was dead. I took the other vehicle and a spare battery (and some provisions) and headed out. I had to pass the security for the power dam being built; they believed the story about a broken down truck. They also warned me that the army was out and nervous about anyone else in the vicinity of the Guatemala border.

At the turnoff to the CCDC ranch, two fellows with a rifle were waiting. I passed out bananas, orange juice, and peanuts. We left the black Toyota without 4WD and started walking the three-kilometer trail through the jungle. One started off carrying the battery on his shoulder. He lasted about 200 meters and then passed it to the other fellow who passed me the rifle. There I was, treading along a jungle pathway, wary of Maya site looters, timber thieves, and nervous army personnel, keeping my finger away from the trigger.

After another 200 meters, the guy (I never was introduced) passed the battery back to the first fellow. A bit further on I figured I would rather carry the battery. I lasted about 1000 meters and then handed it back. He took it 400 meters.

The two Field Officers were there to meet us and took the battery. They put it into the 4WD, and it started okay. There were no chickens; the place had been ransacked. We squished back out through the mud (which is why they could not push-start it). They dropped me back at the black Toyota. They headed to Cayo.

I took the opportunity to see how the dam was coming along. The dam was being built in a deep beautiful valley. There were many twists and turns in the road as it winded down to the Macal River. At one corner I was rewarded with a distant view of a waterfall.

Construction workers give me weird stares. I did not linger to contemplate the ecological damage. Twelve miles down the road I caught up with the white Toyota. Soldiers surrounded it; they had located the rifles. The Field Officers were taken into custody for having firearms without a permit. The 4WD would not start. They were allowed out of the police cruiser to assist.

As we raised the hood up, to check the battery posts, one Field Officer passed me shotgun shells (and I thought they were just rifles). The 4WD started; The Board Chairman drove it as we followed the police vehicle to the Benque station. At the police station, the police searched the two suspects. I figured it was a good time to quench my thirst and excused myself to go buy a soft drink (and dispose of the shotgun shells).

Postscript: *Mid-August 1993 - The arresting officer had been stabbed to death after catching a Guatemalan trying to escape custody.*

AMAZING PLACE

August, 1993

Esperanza, a village of about 250, is three miles east of San Ignacio. Other than five churches along the highway, there is nothing immediately notable about the place. But behind some bushes and old school buses was what looked like an old three-story apartment building. You can find things there you would never think of looking for.

Fifteen years ago Herschel built the structure for his wife and family. Along with him, he brought a truckload of damaged goods. He had been in the business of hauling away damaged or dated goods. He made money for removing the stuff and was able to resell it in Belize where regulations do not prohibit selling damaged or dated products.

Thus he has 50,000 gallons of paint; at $20 a can that is a million-dollar inventory! He has over a thousand windshields, more than all the retailers of Belize put together. He has exercisers, discarded tools, seat covers, oil, shoes, mufflers, soap, coffee, eight-track tapes, carpeting, fixtures, odds and ends, ends and odds. The more you buy, the less you pay per item. You name it; he probably has it or had it. And more is coming. With the flooding in the US Midwest, he has four truckloads of water damaged goods heading this way. The stuff is kept in his "complex," a combination apartment block, and barn and trucking depot. Kitchen sinks hang from railings; boxes are stacked in hallways where items are piled.

I "happened" across this maze of conglomerate production when I was helping paint the (Canada Funded) SIWA building. A field officer was helping the group make it ready for the High Commissioner's official opening at the end of August. I was helping with putting in a sidewalk. They were short one gallon of dark green paint for the outside. We went to this place to see if they could help. Herschel came to the rescue. He asked what colour we wanted. We had not brought a sample but pointed to a piece of carpet. He mixed green with yellow, with toner, with black. Six gallons of toner later we had our gallon.

Back at the SIWA site, the Field Officer mixed that one gallon of enamel paint with one gallon of polyurethane. She got two gallons of glob. I took the "batch" back to Herschel, and he added toner.

It took seven gallons of toner to get those two gallons of glob into nine gallons of paintable solution, but Herschel did it! I took three; Herschel kept six. It took three hours to acquire the paint, one hour to apply it, and seven hours to clean up.

Postscript 2019:*Herschel is still selling items from this location. In August of 2018, I had stopped for breakfast at a location in Cluny, Alberta (population 150). Only one other person was having a coffee there. He knew Herschel.*

WHERE'S THE MOMMA?

September 1993

The farm had half a dozen wood buildings with thatched rooves. I lived in the small cottage and used the thatch-covered area for my kitchen and shower. Having only seen green speckled racers, a non-poisonous garden snake, I had figured that my living accommodations, at the farm, were rather secure. Then the other night while I was walking back to the kitchen (situated under a large thatched roof, with no walls, between the main house and my cottage) I noticed a curled up brown leaf that looked like a small snake, on the pathway. Before I moved it out of the way, I shone my light on it. It was a snake, a dead snake.

At first, I thought it was the cats that had attacked it, but they had not budged from their food dish. Apparently, I had stepped on it on the way to get the flashlight. A few minutes ago I had been walking around in bare feet, preparing for my evening shower. Whatever type of snake it was I had been more harmful to it than it had been to me.

I put the twelve-inch specimen in a jar and the next morning took it into Maya Mountain Lodge where they had a book on Belize's snakes. As I turned the pages, I found a description of one that was pretty close - known as a cohune tommy-goff. In most countries, a tommy-goff is a deadly poisonous snake, but in Belize the locals call most snakes tommy-goffs (to be on the safe side).

The cohune tommy-goff is just a friendly innocent- type snake that happened to have similar markings (triangular head, narrow neck and cat slit eyes) as a fer-der-lance (which is a genuinely poisonous tommy-goff). My sample did not quite match the drawing in the book. I kept going through the book. When I reached the fer-der-lance page - bingo! – it was a perfect match right down to the yellow "beard", known as the deadly *berber amarillo* (yellow beard in Spanish) and yellow-tipped tail. "Oh great, " I thought as I read on. It went on to describe how even the young have venomous fangs from birth, hang out in trees as well as low brush and come out at night or can be seen during the day.

This was a 24 hour, all-terrain, heavy duty, poisonous snake. Its bites are known to kill horses, but if you leave it alone, it will politely go on its way. However, if you wave a machete at it or wildly gesticulate, it will strike. It also said that babies come in batches of sixty. The baby snakes do not carry enough poison to kill an adult. I just wanted to know "Where is the Momma?"

Postscript: *I did not see any more fer-der-lances at that farm, but I came across a huge one, at the next place I lived, one morning while walking along the pathway. I left it alone, and it went politely on its way.*

AFTER MIDNIGHT

Saturday nights I usually watch a couple of movies at Chuc's, where they have some movie channels. They provide the television, and I bring popcorn and soft drinks. Last Saturday, as I tried to sneak home between thunderstorms, I noticed a station wagon sunk on the road, when I was crossing over Higher Street (above High Street and below Highest Street – for true).

Road crews had been working here for about a month. They had left open a two foot deep, four foot wide, twenty-foot long trench, across the road. There used to be a warning sign posted, but it had been knocked down. There was this tail end of a big "ole" station wagon stuck in the trench with four guys standing around cussing. I thought maybe the driver had backed-up as I could not imagine how the front wheels had made it through.

After exchanging greetings, I learned that they had indeed driven through front first. The harmonics of the bounce of the front suspension and forward momentum had been enough to carry the front end over but left the rear end stranded. The cast in this scenario included:

The Driver, who kept getting out to look, and then going back into the car running the engine at high revs, hoping that would get the car out;

Standby, who just stood by mostly watching;

At the Jack, who manned the jack on the rear bumper - alternately putting it up and down (up when the driver was behind the wheel and down when the driver was out);

Wheelman, who tried to change the flat tire with a spare that did not fit; and

Me, who happened along.

A streetlight nearby provided a bit of illumination; heavy clouds dribbling down obscured the full moon.

Wheelman:"*&*^%^*Z^%Z$#%*(*&*&^#@?jack it up (**&%&((&^&$% Just jack up (&^*(&just back the $#@#!@#~*&*thing up.*&*(^^%$^"

At the Jack: "It's going to tip over; it's going to tip over."

Wheelman: "*&^%^&$ Nevermind*&^%^&*just jack it the &^&^$%$% up*^%$%"

Driver: Gets in the car and revs up the engine.

Wheelman: "&^^%><?* Stop &^#&*%$$#%%* Stop *&*&%$#$%Stop!"

At the Jack: "It's going to fall, it's going to fall."

Wheelman: "Just jack *&&^%$#$%^%$ it up" as he finally freed the flat (torn up rubber). This remnant wheel rim was shoved under the right rear fender.

Standby: "Put de rim under de frame."

Wheelman: "*&%& Just &^%$%^ jack, man *&%$#" as he struggled with the spare.

At the Jack: "It's gonna fall, gonna fall."

Wheelman: "*&^%$%* as he still tried to fit the spare.

At the Jack:"...gonna fall" Sure enough the car falls... crumpling the right rear fender upwards and inwards.

Wheelman:"*&^%^$*&?##$$&@*&^@*&..."

Driver:"*&^%^$*&?>$%$#$@%#@@*&..."

At the Jack:"*&^%^$*&?>%&@#$!!&*@@*&..."

Standby: So only I can hear, "You want to buy us some beer? Maybe you're thirsty?"

Me: "No thanks, I'm doing fine."

Wheelman: "Jack de *&%$&&%$&#$@@&% car &%$#*%^#@!."

At the Jack: "It only going to fall."

Wheelman: "Jack de car *&%$*&#$!@&$@*&^"

I helped to steady the car. After several revvings of the engine and an uncountable number of cusses, the spare was finally shoved in place. The lug nuts did not fit right; two were lost in the mud. The town power failed and the light from the nearby streetlight no longer illuminated.

Stand-by: "How about a couple of bucks for a beer? You got a couple of bucks?"

Me: "Yes I got a couple of bucks, but it is not going to be for a beer."

Driver: After revving the engine the rear wheels spin in midair. "It no go."

Wheelman: "Course it no go &**&%#$%@$^#!@ ain't got nutting to go on &*%$#!"

At the Jack: "We got to lift, got to lift."

Wheelman: "Need a *&%#%$@!$% plank &%$#*&%$ under de wheels *%$#$"

Driver: "Push you di hard" as he gets in and restarts the engine.

We push di hard but it no go.

The flat under the right rear smoked rubber as the wheel spun on it.

Standby: "You thirsty now?"

Wheelman: "We got to put some rocks under de wheel" he gathered some rocks.

I helped. After five or six rocks the driver tried again. We pushed and lifted as the tires spun in a high whine.

Wheelman: "*&*%$#$!#%$&*&$@!%"

Driver: "Me make it go fast."

At the Jack: "Me jack it up for more rocks."

Standby: "Just a couple of bucks for a beer."

Wheelman: "*&%$#!@#$*&%$ Just lift harder &*$#@#$ push &*%$#"

Me: "No I'm not giving you a couple of bucks; what about these guys?"

Standby: "Oh no, just me man."

Wheelman placed more rocks under the wheel, and At-the-Jack put it higher. The driver got in and revved like crazy. We lifted, pushed and grunted in exertion. Wheelman swore and cussed. Standby kind of leaned against the side of the vehicle. It no go. We yelled at the driver to stop.

Standby: "You thirsty now?"

Me: "No … and anyway you guys should be buying me a beer."

As I gathered more rocks and piled them near the car Wheelman started putting them under the wheels. I kept bringing them. They stood behind the car again as the driver started it. Even Standby pushed (a first). This time it exited the trench. They piled in and offered me a ride. I declined but advised them to have the flat fixed. After saying goodbye and heading on my way, I heard a faint plea from the backseat: "Two bucks, just two bucks." The rain increased. The street light came back on.

THE CHOICE

Early September 1993

The President for the Crystal Hatcher's Association, went to Belize City to pick up eggs for the hatchery project. The Field Officer who had been making the run was trying to get the group to do more of the tasks including the cumbersome customs paperwork.

The camper, on the Black Toyota, (purchased for helping with the picking up of eggs and delivering chicks) was not staying on very well due to the bumpy roads. Two weeks earlier we had tightened down the camper with double and triple nuts.

About 3:00 p.m. on that pick-up day, someone had dropped off the cartons of eggs at the CCDC office. There had been an accident, the rear tire had blown, and the truck had flipped over.

After examining the eggs, not being able to see a cracked one never mind any smashed ones, I concluded they were exaggerating about the accident. The vehicle could not have turned over, or it would have made one big omelet. We're talking 2500 eggs.

The Field Officers fired up the infamous White Toyota (which meant tracking down a charged battery first as the alternator does not work) and delivered the eggs to the hatchery in Bullet Tree Falls, leaving me on my own at the office.

Just as they were pulling out, the eldest daughter, of the President of the Sewing Group, came in all upset. She reported that her mother was in the hospital with a broken arm and I should call her. "Everything happens at once," I thought. I did not think there was a chance of speaking to a patient at the hospital, but they put her on the line for me. It turned out she and her youngest daughter had been in the back of the Black Toyota when it had turned over. The baby was okay. As I talked with her, a distraught woman came in telling me about the poor fellow still being at the wreck and how he was hurt the worst. She was his wife, and she too had been in the vehicle with their baby. She got off with a lacerated leg. The story began to come together.

While going around a curve, the driver's side rear tire blew. The vehicle went into a slide across the highway heading for two women, walking along the other side. He had turned the wheel sharply causing the truck to flip.

He had the presence of mind to get someone to bring his passengers and the eggs to Cayo. He stayed with the vehicle even though he was soaked in blood.

After delivering the eggs, the Field Officers proceeded to Ontario Village to inspect the damage. They asked if I wanted to come along, but I declined; I was confident they could handle it. I did inform the police and left a message with the Chairman. The Coordinator had the day off.

I thought this would be a story I could tell without being involved, but they were back after getting a ride with a sympathetic motorist. He dug out the heavy-duty chain, from the back, and would use the motorbike to take it to the crash site.

They planned to tow the Black Toyota back with the White Toyota. That meant they would have three CCDC vehicles there but only two CCDC staff to operate them. I was not fond of the motorbike; I had an "incident" with it. Besides having to be push-started, it had a broken clutch. I had run into a cement wall once while trying to avoid a vehicle when the throttle stuck. I rode on the back with 100 pounds of chain in a backpack.

At the crash site, the Black Toyota looked like a vehicle that had flipped over. It had been righted by some passersby's. The windshield was smashed, the roof was crumpled in, the fenders all had creases and dents, the hood was folded, and the camper had come off, but the double and triple nutted-bolts were still in place.

The camper aluminum lining had simply ripped off around the nuts and bolts. We removed the camper and placed it in the White Toyota. The Field Officers replaced the blowout with the spare. It rubbed against the fender and was low on air. We had three vehicles to bring back, and now there were three of us.

The Field Officers were gracious enough to give me a choice. Behind door number one, should I choose it, was the rescue vehicle: the White Toyota with four bald tires, a battery that might conk out, a bad tie rod end and a camper in the back that blocked the rear view.

This vehicle would be pulling the Black Toyota. Behind door number two, was the Black Toyota, with a smashed windshield, glass fragments all over, and crumpled-in hood, and shaky tires. This was the damaged vehicle to be pulled by the "rescue vehicle". Behind door number three was THE bike (the clutch was broken, and the throttle was known to stick). Have you had times when you did not want to be responsible for making the choice before you?

DAMN

The Belize River may be one of the last unpolluted rivers in the world, according to an environmentalist who recently gave a talk at the San Ignacio Hotel. He said the present dam project stands to not only take away that distinction but to turn the Macal River into a half year mud trickle with assorted environmental devastation.

The speaker told us that a CIDA (Canadian International Development Agency) funded study was undertaken to determine the potential impact of the dam. It looked at two of the proposed four stages, but the recommendations were not being followed. The powerhouse and generators, as being built, will require all of the Belize River water. To achieve this, a series of dams are to be constructed accumulating in a huge "fourth" stage reservoir fifteen miles long. This is huge by Belize standards.

Large tunnels are being built to direct water from one dam to the next. There will be very little water between dams leaving stretches as long as eight miles that will become little more than mudholes - an unpleasant sight for tourists, a dead zone for the environment.

Moreover, the dams themselves will not be filled year round. There will be cyclical periods of low water levels leaving a barren, unsightly zone of rock and mud. The turbines will operate to meet daily peak needs, meaning the Macal River will experience daily flow variations. There will be times of low generation needs when the river will be shallow (unnavigable) and times when the river will run quickly (unmanageable for boats).

Moreover, the water being released, after turning the turbines, would not be suited to river-life, as Belize wildlife requires. The water would be alternately hot and cold. After sitting in a dam, water tends toward cold and after running through the turbine, it becomes heated. Either way makes it sterile as a life supporting system. These daily flushes will be detrimental to the food chain.

Another danger from tropical river banks being exposed is the possibility of mercury combining with elements to make a poison more deadly than dioxins. Other effects to be concerned with are the loss of the annual flood renewal factor, leakages into other drainage systems, swamping of caves, loss of recreational provisions (swimming, bathing, boating, etc.). Mud being released could lead to permanent damage to the reef.

Belize's special wildlife is in danger from the dam. To mention only two: the Scarlet Macaw and Tapir, stand to lose their habitats depending on how far the development occurs. The dam project is not likely to stop at the smaller dam components, which will leave the equipment incapable of operating at full capacity and therefore not meeting electrical demands.

Postscript *2001: They are planning to go ahead with other stages flooding greater areas. Jonathan Lohr is assisting with the environmental cause.*

Macleans Magazine published an article (August 27, 2001) on the proposed project.

Postscript *2019: See the Epilogue for consequences of the Dams*

REVENGE OF THE KITTENS

September 1993

Catsup had kittens. It was only a week earlier that I noticed that she was pregnant. I don't know if she was that way when she adopted me or if the black tom cat that had been lurking around was responsible. Either way is conceivable. The morning of the day she gave birth she was acting strange, meowing a lot and not eating. Tobie had behaved unusually the day she had her puppies.

Catsup even showed me her "nest". She had picked a spot on one of the upper bunks where there was a large piece of foam. She nestled in the "tunnel" where the foam was folded. I left her as I headed to work. Sure enough, that evening when I came home, there were four little kittens.

WEEK ONE: Three of the kittens had the same orange and white colouring as their mother but with a preponderance of orange. The fourth had that colouring but also some black splotches - known as a calico. The kittens were no bother, made no noise, and mother attended to them most of the time. Sylvie was puzzled as to why she did not have a playmate (Catsup) to torment anymore.

WEEK TWO: Every once in a while one of the kittens somehow made its way on top of the foam and made squawking noises until I moved it down again. They all seemed to be female. They have revealed their baby blue eyes.

WEEK THREE:The kittens exhibited different behaviours: Bogart liked to curl up in my hands, Cagney liked to explore along my arm, Raft whined when I picked her up, and Edward G., the calico, displayed all three temperaments.

During week three I came down with Dengue Fever. Mosquitoes transmit this bone crusher disease. For the first week, I was incapable of moving; I had a high fever.

Rolling over in bed required great effort. Getting up to visit the bathroom was agonizing. I did not eat. When I had not shown up at work for several days, two CCDC Filed Officers came to investigate. They took me to the

Doctor, but the fever had broken by then. When the Doctor took my temperature, it was only 103. Still, he gave me a shot to bring my temperature down. After a week the fever was gone, but I was very weak.

I went to work for only two or three hours a day. I experienced severe lethargy for three weeks. Fortunately, someone told me that this was part of the recovery; that helped me deal with it.

I moved to a bed in the main house. Catsup chose this time to move her kittens to a box beside my bed. They at least looked like kittens now. She did not move them all at once. She placed Bogart in one box and then a little while later Cagney in another. An hour went by, and she had not brought anymore. (I'm stuck in bed and only capable of watching.)

I think she was putting each kitten in a different place. The next time I was able to work up energy I checked the next room; Raft and Edward G. were still there. A half hour Catsup brought the others to join Cagney and settled down. Four more hours went by. Bogart was still isolated and still sleeping. Either Catsup forgot or cannot count. I moved Bogart over. Catsup lifted her front paw in welcoming, as if to say, "Oh there you are".

WEEK FOUR: The kittens wanted out of their box; they meowed regularly. When mother attended, they just wanted out of the box. I let them investigate their surroundings. I was unsure if they could be left alone. Hopefully, their mother would return them if they got out of hand. Catsup ignored them. After some indecision I returned them; I did not want lost kittens.

Second Day: Their box was moved to the kitchen. I took the kittens out more often. As soon as they are out, they follow me around as fast as they can scurry and scamper. "Hey guys, I'm not the Momma."

After two more days of in and out of the box, I relented and turned the box on its side so they can leave and enter at will. Everywhere I step I need to look carefully. Sylvie has taken to picking on the little gang. I suppose it is playing, but it seems on the rough side, but not as bad as Sylvie was with Catsup. The kittens did not seem to mind, but they are not reciprocating. Amazingly, they returned to their box and spend most of their time sleeping.

Fifth Day: The kittens began licking themselves and making faint striking motions with their paws. Mother ignored them, especially when Sylvie was attacking them. When Sylvie and Catsup were eating, the kittens gathered around the feeding bowl and interfered with the intake. Catsup and Sylvie did not know what to do; they cannot eat in peace. The kittens were getting their revenge for mother not playing with them and for Sylvie being too vicious. I look forward to the day when the kittens can gang up on Sylvie.

Sixth Day: I started training the kittens to drink milk from a bowl. On the first attempt, only Cagney would lick milk from my fingers. Two days later I made another effort. Cagney and Raft drank from the bowl by following my milk-dipped finger down, but I still had to dip their heads towards the milk. Edward G. licked milk off my fingers but not from the bowl. Bogart would not indulge. I wanted them to feed independently before taking their postings. Nevermind.

WEEK FIVE: It seemed every time I offered the kittens a bowl of milk I had to retrain them how to drink. Bogart had not caught on yet. They began swatting back at Sylvie, but that just made her combat harder and made them cry out. Mother was demanding on me, wanting to be fed all the time. The kittens entered a frisky stage: running every which way; hiding in nooks and crannies and exercising their claws. I hoped to farm them out soon.

WEEK SIX: It was close to decision time. Milli (a member of the Leatherworking group) wanted three kittens: one for her daughter (age 5), one for her nephew (6) and one for her niece (2). Which three to give them?

D–DAY

October 24, 1993

Delivery Day - Bogart still had not got the hang of drinking milk out of the bowl, so that meant Cagney, Raft and Callie (the calico) were candidates to go to Aimee, Jason, and Val. I packed milk, cat food, flashlight, rain poncho, and water. I decided to carry one kitten in the pack and two in my arms. Callie did not mind the pack.

As I started with two in my arms, I knew this was not going to work for five miles. My arms tired. I hunted up a box. Living away from everything one saves everything. Without garbage collection, anything I wanted to discard had to be burned, composted or buried (so I tended to purchase things that did not require disposal) consequently I had boxes, bags, and jars galore. I figured the box, the kettle came in, would do just fine. The kittens were made cozy with some padding in the bottom. Every few hundred yards I would check on them, and they stayed fine. At one point Callie and Cagney wanted out.

I positioned one on each shoulder; that suited them. Raft still liked being carried in the box. About a mile from Milli and Tina's place Raft wanted out too. I placed her on my hat. When I showed up at the delivery destination, the kids ran to let me in through a large gate into a large, well-kept yard. They knew for days beforehand that I would be bringing the kittens but were not prepared to see them riding on my hat and shoulders. They were quite awed by the spectacle. Milli laughed and clapped her hands.

At first, the kids were afraid to hold them. Then there were not enough to go around. The poor things had tears in their eyes as they clutched at me in fear of being hung upside down. When a child came to take a kitten away, it clung to me desperately. Milli made a "house" out of two boxes, and the kittens were placed inside. The rule was that the kittens could not be taken from the box. That was their refuge.

NUCLEAR TESTING IN ALBERTA

October 1993

Why am I writing about nuclear testing in Alberta when I am in Belize? You will just have to read this eclectic story. It had been about a year and a half since I had been to the Cayes and I needed a break. I was getting over Dengue Fever, and some beach and sun would help my recovery.

At Caye Caulker I checked into one of the last rooms in the "hotel". Then I went to the 'Cut' at the north end where you can suntan and snorkel. There were quite a few changes since I was there last - about 50% more "hotels" and restaurants. The thatched hut at the Cut had been trashed from Tropical Storm Gert (before it became Hurricane Gert) and the dock was listing inland. This was explained when I attended the evening slide show titled "Reef Ecology".

There were plenty of people sunning on the tilted dock, and I stepped over some bodies to a vacant space, greeting people who bothered to notice my passing. Wait a minute; I'm sure those last two women had no top on! I refused to steal a confirming glance; females ought to be able to obtain a suntan topside without being ogled. I could be nonchalant about it. Undoubtedly they secured their share of unwanted attention.

Sure enough, some fellow came over pretending he was looking for something. After he left, I overheard them speaking German, or I thought it was German. After getting my goggles and snorkel adjusted, I looked for the ladder and realized it must be on the other side.

I was not going to let any bare-breasted Germans intimidate me so I stepped over them but still could not find the ladder. It must have been damaged when the storm hit. Nothing to do for it but tumble in and readjust the aquatic gear as necessary.

Snorkeling offers some underwater viewing opportunities. The amount and diversity of fish has been less with each visit. It may have something to do with seasonality, but I think there is an inverse correlation with the number of speed-boats shooting through the Cut. They traversed through, about every eight minutes, whereas the first time I was there, I only saw two speed-boats.

There were some interesting fish, starfish, conch, huge schools of guppies (what do they eat? and what was feeding on them?) and then a four-foot stingray loped through the area. The Reef Ecology slide show was very informative. It started out pretty boring: "Here is a picture of part of the reef, outside the reef is the deep ocean, inside the reef is shallow water, see the waves breaking against the reef.

The Belizean Coral Reef is the second longest in the world. Australia is first at 2,070 miles. Belize's is 180 miles long."

The slides showed the marine life, "who eat what when"; colour changes; how female fishes turning into pseudo-males; why brain coral looks like mountain ranges; day feeding coral is coloured but night feeding coral is grayish; how cleaning shrimp chase away the bristle worm who eats the toxic coral (which deters other creatures) to give itself toxic protection; how cutting down the mangroves (which shelter tortoises, manatees, sea horses, and provide breeding ground for fish) leaves the Cayes unprotected.

Although there is breeze most days, when a tropical storm hits or hurricane comes close the wind does no damage, but the water rises over the island and then when it recedes it takes away part of the shore - that was why the dock, at the Cut, was listing inland. The part that had been anchored on shore no longer had a shore to be attached to. Humans are a great hazard to the reef.

Next morning I went for breakfast. All the tables were taken, but there was one guy sitting at a table for seven. I asked if I might share his table and he introduced himself as Paul from England. Another short-haired chap, John, joined us, and I realized they were two of the British soldiers stationed in Belize. They were entertaining enough. John kept ribbing the waiter "hey we've only been here half an hour and no coffee yet."

"It's coming, it's coming" replied the overworked fellow. "Now don't get fresh," says Paul. "The only thing that's fresh" retorts John "is that coffee that's not here". Finally, the eggs and toast arrived ($3 for the toast and $1 for the eggs), but there was neither butter nor marmalade as promised on the menu. Hash browns were not available either.

While John was making some jokes about having to catch the cow to get some cream to go with the coffee, Richard and George showed up.

It was now 11:00 a.m.; it had taken one hour to be served. George and Richard were not going to wait for coffee; they started on beers. The others groaned. By noon they were on rum and cokes. I had one too. Asking some polite questions, I enquired where they were from and where else they had been stationed. They talked about many things.

I learned about regimental holidays, traditions, exam procedures, Gurkhas, British military humour, regimental badges, old timers stories, regimental solidarity, the Gulf War, the Falklands War, postings in Australia, Cypress, Germany, Kenya, poison gas testing, bacterial warfare testing and a nuclear blast at Suffield.

These guys riddled off one anecdote after another. They had been talking about General Powell retiring and the gay issue. They told me their position. "First you find 'em then you trash 'em, and then they are out," was the reply.

"What's your guy's personal position?" I enquired. They continued talking about not wanting to have to shower with a gay person, nor with a woman, (trying to be fair?). They went on at some length about showering and then Richard provided insight by saying to George, "You know, I'm with you on this George, but the problem is us. Here we are worrying about showering with a gay or a woman, and for all, we know they couldn't care less about showering with us."

I was on my fifth rum and coke, and Richard was talking about his time served in Suffield, Alberta. His fondest memory of Alberta was Electric Avenue. "Oh, wouldn't they let you into Medicine Hat?" I jibed.

He told me how he was in Calgary when the Flames won the Stanley Cup. The party on Electric Avenue was the biggest celebration he had ever attended. It was greater than any European soccer win, bigger than the Queen's 25th anniversary, bigger than any regimental bash - so he said.

Richard was just warming up about his time at Suffield. It's funny what some people will tell you without asking. I was not surprised about the chemical testing, and I was not even surprised about the biological warfare testing (though it made me shudder), but when Richard talked about the nuclear blast I was attentive but trying to appear nonchalant (good thing I had some practice the day before). There were dozens of questions going through my mind. He talked about the crater, the two mile-zone, the three-mile zone, the five-mile zone, the seven-mile zone - things you associate with an atomic explosion.

He talked about observing radiation effects on test monkeys. There was one with a metal box on its head. He then mentioned about sealing up certain places that they did not want the Russians to see when they came to inspect (!?!?).

I thought maybe it was a dynamite equivalency test, but why the experiments with the monkeys and inspections by the Russians? Richard was posted in Suffield for six months in 1989. I wondered when this blast was to have occurred. Then George talked about women sunbathing nude in

Germany (I told you those sunbathers were German speaking).

Maybe Richard was getting mixed up with testing in Australia? They had atomic explosions there and in atolls in the Pacific. Had I missed something in the news? Is this common knowledge or something we are not supposed to know about? Humans are the greatest hazard to the environment.

Postscript: *After sending this story to friends several of them made an effort to determine if there was any truth to the testing, including writing to Canada's Minister of Defense. A reply was received denying that any nuclear testing had been done at Suffield.*

A CASE OF MISSING BUTTERFLIES

November 1993

"Dr. Pat" and I went to Punta Gorda for the Garifuna Day long weekend. Pat sold me on the idea by saying that it would only take six hours to travel the 160 miles. Michelle, another Cooperant came along. It was quite a ride! The road steadily deteriorated.

The first thing to fail on Pat's van was the brakes on the left front wheel. We had noticed a burning smell, and another passenger spotted smoke coming out of the driver's side. While we looked under the hood for a radiator leak, then someone noticed smoke coming from the wheel. Without touching it, we could tell it was very hot. After prying off the hubcap, we could see the rim was all black and seething heat. Pat had just installed new disc brakes. We figured that there had been something wrong with the installation. Another consideration was that Pat had been riding the brakes all the way through the rough part of the Hummingbird highway.

We let it cool and even took the wheel off but detected no abnormality. We postulated that it was just a matter of breaking-in adjustments and carried on with periodic checking. We missed the next turn off and wandered into Dangriga and out again before finding the Southern Highway. The dirt road was badly pot-holed; it was slow going and was now dark. It started to rain. Somehow we missed another turn-off and found ourselves in Independence. That was probably a good thing as the left rear shock came loose.

Fortunately, we found a mechanic's shop. He was out at that moment, but I borrowed some of his tools to remove the shock. He accused me of doing mechanic duties (for removing a bolt). The shock was not reparable, but at least it would no longer be dragging and bumping.

The road worsened, and we became stuck in the mud. While pushing to get the vehicle out, I performed a belly flop into a puddle. It took us six hours to complete the last forty miles. Fifteen miles before Punta Gorda, the bridge was closed; we had to take a twenty-mile detour. We straggled in at 3:00 a.m...

The others had friends to stay with. Pat and I ventured to Nature's Way Guesthouse. For the rest of the weekend, I was not inclined to go anywhere that required conveyance by vehicle. I hung around the guesthouse and walked around town.

Punta Gorda is a friendly, seaside town of about 3,000, a nice place to live. It is the jumping off point, by ferry, to Guatemala. There are even some restaurants, but they were mostly closed on the holiday.

Nature's Way Guesthouse staff were away. Thus when the phone rang, I explained to the caller that the Manager was helping at a workshop. He pleaded to have the Manager call back. Recognizing the voice as Mick Fleming, the Co-owner of Chaa Creek Resort., I inquired if I could help.

After identifying myself, and explaining that I was taking some time off, he explained his crisis. Two of their guests had mistakenly taken a case of rare butterflies meant for an overseas collector. As far as he knew, the elderly couple was heading to PG and could I try to locate them? The Chaa Creek naturalist had gathered the half-dozen butterflies over a period of several years specifically for this collector. They were very valuable.

The evening before I had seen dozens of foreigners "out and about". How was I to locate the right two? I decided to follow two strategies: first I would inquire at the twelve hotels in town, and then question every elderly tourist I encountered.

The first hotel had its gate locked. I wondered how guests registered. The second place was a combination hotel transport depot. I was told I could ship a parcel but not leave a message. At the third place, the receptionist could not grasp the concept of leaving a message for someone who had not yet arrived. Tucked off to one side there was a grungy place called Butterflies Inn. Although the name was enticing, it appeared to be no more than a two-table food outlet. I gave it a miss. That was it for the south end.

As I headed to the other side, I pondered this mission: 'What difference did it make who had the box of butterflies?' They were no longer in the forest. That could be why they are rare. Maybe it would be justice not to reward the collectors with the return of their butterflies.

But then I figured the naturalist would only be requested to collect more thus further depleting the supply. But then again what if the collector shows up in Britain and shows off his samples and stimulates a demand? Then maybe he would not share his source. Such deliberation under a noon sun - I opted for an ice cream cone.

Leaving the ice cream stand was an elderly tourist woman. Following my second strategy, I inquired if she was a former guest of Chaa Creek. She was astonished at being so identified but replied affirmatively. I explained the case of mistaken butterflies and arranged to retrieve them on Monday. She was just getting on the once a week bus heading out to one of the villages and could not take time to give me the butterflies.

My strategy of asking every tourist I encountered paid off. Had I been two seconds later, stopping at the "Butterfly Inn," she would have been on that bus and gone. It was just a case of missing "Butterflies/Butterfly's".

There is a triple pun in the title: 1) A Case of Missing Butterflies where the Case is an actual case; 2) where the Case is an Incident, and 3) where Butterfly's the place I missed going to and thus catching up with the couple with the Butterflies.

Postscript: *In 2000 Michelle was killed instantly by a drunk driver after she stepped off a bus in Punta Gorda. The driver was not charged.*

Postscript *2019: Mick Fleming did not recall the incident when I mentioned it to him.*

HOME LEAVE

December 1993

Friends from southern Alberta brought some much-appreciated jars of homemade strawberry jam. They were going to stay at my place over Christmas while I travelled back to Canada. Normally I would have hidden away my stove, gas cylinder, solar panel, and other goodies, but I left them out for my guests to use.

I had been in Belize for over two years and had completed my contract, which was renewed, and I took vacation time to be home for Christmas. I used the last of my money for a cheap flight out of Cancun. I still needed to take a bus from Calgary to Vernon, where my parents lived.

Population information

With an estimated 390,000 people in 2019 and an area of 8,800 square miles, Belize has a population density of 37 people per square mile. This is the lowest in Central America and 177th in the world. For comparison, Canada, with thirty-seven million people, has 11 people per square mile and the United States, with 330 million people, has 93 people per square mile.

Population by decade (to the nearest 5,000)

1950	70,000	1960	90,000
1970	120,000	1980	145,000
1990	190,000	2000	250,000
2010	320,000	2020*	400,000
		*projected	

Ethnic Groups

The Maya have been in Belize and the Yucatán region since the second millennium BC. Belize's original Maya population were devastated by prolonged drought and conflicts between warring societies. Many died of disease after contact with Europeans. Three Maya groups now inhabit the southern part of the country: The Yucatec (who escaped the Yucatan Caste War); the Q'eqchi' (who escaped enslavement by Guatemala) and the Mopan (indigenous to Belize, forced to move to Guatemala by the British and returned).

The first Mestizos (of mixed Maya and Spanish descent) came to Belize, to escape the Yucatan Caste War. Mestizo refugees also came from El Salvador, Guatemala, Honduras, & Nicaragua during the 1980s due to conflicts in those countries. Mestizos account for half of the Belize population.

Creoles/Kriols, account for one-fifth of the Belizean population. They are descendants of the Baymen slave owners and slaves brought to Belize for the logging industry. These slaves were of west and central African descent from other British colonies.

The Garinagu (Garifuna in the singular), at around five percent of the population, were either the survivors of shipwrecks or took over a ship they came on. Based on a genetic study, their ancestry is three quarters Sub Saharan African, one fifth Arawak/Island Carib with some European. Punta, distinctly Afro-Caribbean, is a popular genre of Garifuna music and is popular throughout Belize. Dangriga, a mostly Garinagu community, has a yearly Easter fishing tournament. November 19, 1832 is "Garifuna Settlement Day".

Although only composing three percent of the population, <u>German speaking Mennonites</u> make a significant contribution to the economy. Before the Mennonites arrival, Belize imported 80% of its food. Now it is less than 20%. Most were of 18th Century German ancestry that migrated to Russia. Their descendants moved to North America. Some reached via Mexico in the 1950s, when Mexico wanted their children to attend Mexican schools. They live in communities like Spanish Lookout and Blue Creek. There are also Old Order Mennonites from the United States and Canada that arrived in the 1960s. They may look similar to the Amish, but have differences in their interpretation of the bible. Some projects I assisted with were agricultural related, trying to break into the Mennonite market dominance.

1994 – THE LATE CLASSIC PERIOD

UP THE CREEK WITHOUT A PADDLE

My visit back to Canada was enjoyable, but I had enough of snow and cold. I was returning to my jungle paradise.

ROBBED

January 1994

On the same flight to Cancun were two members of the Calgary Rain Forest Action Group. They were planning a two-month tour of Central America. I told them they would be welcome to stay with me during their time in Belize. In Cancun, I met up with friends of my sister, who were going to stay at my place while investigating the possibility of establishing a winter retreat in Belize. On the bus, to Belize City, we met a fellow from Vancouver who I invited to stay also. Even a fellow driving a fire truck from Edmonton to Belize would look me up. Yes, everybody had a place to stay.

"Had" is the key word here. My clothes, books and bookcases were all that was left when I arrived "home". "They" took just about everything else: beds, blankets, sheets, stove, pots, pans, dishes, cooking utensils, cutlery, artwork, Walkman, cassettes, furniture, flashlights, radio batteries, machetes, solar panel, car battery, overhead light and even the dozen jars of strawberry and raspberry jam. Four thousand dollars' worth of stuff! I had not realized I had $4,000 worth of stuff to lose! It is not like I felt I had lost $4,000 worth of goods - it is just that I could no longer accommodate guests.

There are no beds to sleep in, no chairs to sit on and no table to eat from. Even if I had a stove to boil water, there are no cups to hold tea or coffee. There were supposed to be people staying at my place over the holiday period, or I would not have left everything out in the open. Normally when I went away, I took the stove and propane tank to neighbours. The solar panel configuration had been promised to Charlie. However, as guests were staying at my place I left everything out for them to use.

It would not have prevented the thieves from taking the waterbed! I could not believe they made the effort to take apart and move the waterbed and did not take the attachments. I wondered how they would get water back in it. The bed was such a nuisance to move, and the solar panel must have been a hassle to get off the roof. I suppose they had lots of time - they took at least two truckloads - with their organizational, logistical skills these guys ought to be in the moving business.

Manfred was my landlord. His hired hand had been chopping bush there in early January and had reported to Manfred that he had noticed someone sleeping at the farm. Manfred was under the impression that my guests were still there and did not investigate.

The thieves must have been there a few days, waiting to see if anybody would show up before they picked the place clean. Maybe they would not have stayed if the stove and lights were not available?

The only furniture they did not take were the bookcases I had made. Was that a reflection on my carpentry skills? What about my clothes? Is there something about my taste in clothes that did not merit stealing? It was bad enough to 'thief' from me but did they have to be insulting?

My sister's friends stayed at a hotel. On the weekends I introduced them to local resort operators so they might get a feel for what worked and what did not. The Manager from Maya Mountain Lodge told them it would be a tough hoe. The Manager, from the Parrots' Nest, said they needed to know their target market; at Chaa Creek, they were made aware of high level organization; at Blancaneaux they marvelled at the splendour, and other small operators gave them various insights.

One of the things suggested to them was to go into partnership with Belizeans. I took them to meet the Chuc's. Moses had expressed interest in starting a canoeing venture in Bullet Tree Falls where the family has riverfront property. I found out that their eldest son had been killed in a car accident while I had been in Canada. Whereas I had lost some items, and a means of putting up guests, Luis had been robbed of his life and the Chuc's robbed of a son. I may be able to recover my loss, but they will never be able to replace theirs.

A WHEELCHAIR FOR KENNY

February 1994

Two years ago Phil Watson visited Belize. We had worked together in Botswana where Phil had set up a shoe factory. While visiting in Belize, I had him instruct a leather-working course for the Western Leather Cooperative. Phil did the course at no charge, taking a week out of his vacation time. The course went well; the participants went away with enhanced skills and renewed enthusiasm.

While in Cayo I took Phil around to some of the other projects. After a weary day of driving around on rough roads, in a truck with poor suspension (the infamous White Toyota). Phil insisted on a refreshment stop. The closest spot was the corner bar in Bullet Tree Falls. Phil paid for a couple of cokes and asked to be left there rather than ride any more Belizean back roads.

That afternoon we were the only patrons in the living room size bar. It had a colour television, turned to the only channel rural folks could watch. Electricity was a recent addition to the village. Sitting in his wheelchair, watching the television, was Kenny, a seven-year-old paraplegic.

He was dependent on others to move him from spot to spot. His wheelchair was pretty dilapidated. I was not even sure he could follow what went on, on the television. Phil went over to Kenny and started talking to him. He made Kenny smile, which was more than I had ever seen him do on my previous visits when I stopped in for a cool one during softball practices. Phil engaged the father in a conversation about the boy and made some suggestions about how he could move around better.

Phil owns and runs an ambulatory footwear shop in Dundas, Ontario. He looks after people with special mobility needs. When Phil returned to Dundas, he followed up by sending some literature on how to help kids like Kenny. He also kept an interest in the leather group and would always ask about how they were progressing. He did a leather feasibility study for them at no charge. His study was used as an attachment for the project proposal seeking funding for the tannery and leather shop.

When Canada Fund approved money for a building and equipment, Phil was the one in Canada, who secured the best deals on used equipment, packed it in shipping crates and then arranged for transportation. Again he did all this at no charge to the group (but at some expense to himself).

When the shipping crate of heavy stitchers, skiving machine, splitting machine, and lace maker was opened up - there were the parts of a reconditioned wheelchair tucked away in the crevices and spaces not taken by the machinery. When I dropped by to have a look at the goods I was "given" the task of assembling the wheelchair. There were no diagrams showing how A connected to B, but I managed to put the bearings in place; the arms hooked correctly; the footholds placed, and the wheels connected.

The tires were shot. I had to scour both Santa Elena and San Ignacio for 24" by 1 3/8 " tires. The places I visited had 26" x 1 3/8" and 24" by 1 3/4" but not the size needed. One shopkeeper suggested I check with the fellow, who specialized in lawn mower repair, and changed tires, about using an odd size. He thought the 24" x 1 3/4" would work.

He offered to do the installation for free as it was for a good cause. He provided the names of several places that might have the right size. Martha's did not have them; Valdez was out, and Martino's had only the other sizes. Celina gave me 20% off on a pair of 24" x 1 3/4". I left them with Rene and checked to see how the leather group was doing with their building. Back at Rene's, I found out that the tires would not fit, so I headed to Herschel's. Lindsay Herschel had just about everything but no 24" x 1 3/8" tires. While at Herschel's I picked up a stove and propane tank to replace the ones stolen. When I took the stove to my place, I found I had to assemble it.

Egad, two things to assemble in one day! When I was done fumbling with that, I found the instructions taped to the last piece. It was in Spanish anyway. When I tried the stove flames shot out the back. The next day I checked with Valdez.

He took me into a large dingy room in the back of his shop. It was filled with beer, soft drinks, and bicycle tires! Surely, if some place was going to have an odd size, this was the place.

We found one right off! Then it took thirty minutes to find another, with me climbing up in the rafters and Valdez tossing bundles around. I did not think we were going to find another one; they do not get requested very often. One was priced at $13 and the other at $17. Valdez was not giving any deals and did not bat an eye over there being two different prices. I paid the $30 and headed for the lawnmower man. Rene put the new tires on and would not even accept two dollars to buy soft drinks.

After work, I took the wheelchair out to Bullet Tree Falls. Kenny was sitting outside; he did not even give me a look as I pulled up in the CCDC truck.

But when I lifted the wheelchair out of the back his face broke into a huge grin. And when I took it over to him and indicated that it was for him, he actually sputtered out THANKS. He did not talk very often, but this gift had moved him so much that he demonstrated his gratitude with a verbal emotive and a wave of his arm: hidden abilities. I wondered if he knew how to stop stoves from belching fire out the back?

MAKING UP FOR POOR DESIGN

February 2, 1994

Every once in a while I assisted the leather group with the construction of their tannery and leather shop. When they were digging their effluent pits, I spent a half day shovelling dirt. When they poured the concrete floor, I helped mix the cement.

This morning I helped with the rafters. It was poetic justice that I ended up helping with this particular task. The Treasurer for the group and I were fabricating eighteen-foot lengths of 2" x 6"s to go down the sides of the roof. They would be used to support the purloins, that the roofing zinc would be nailed to. There would be a two-foot overhang on each side. We picked out eight-foot and ten-foot lengths and "joined" them. This was not a simple chore.

The 2 x 6's are rough cut and usually warped. It is okay if they are warped width-wise but not length-wise. After sorting out the lengths, we then had to saw the ends off square. After that was completed, for forty-two pairs of planks, we lined each set up to make sure they were straight. Then we carefully set six-foot lengths of 1" x 6" planks across the joint, making sure they lined up with the "top" side; and nailed them. We then turned the assembly over, checked it was straight, and nailed another 1" x 6" on that side.

As we did this for all those pairs of boards, I asked questions. "Wouldn't it be easier if we just had 18-foot lengths of 2" x 6"s?" "Yes, it certainly would," he replied, "but that takes a special order." After assembling a few more of these rafters, I looked up at the crossbeams that would form the base of the rafter 'triangle'.

I knew the building is supposed to be 30 feet wide by 40 feet long (I'll come to why I know that). The crossbeams were also fabricated joints. That meant they were roughly fifteen feet each. I asked Fred about that, and he responded, "Yes, they were made by joining a fourteen-foot length to a sixteen-foot length."

"So 16 foot lengths are easy to come by?" I asked. "Oh definitely," he said. We painstakingly assembled a few more rafters, and my mind made one of those intuitive leaps.

"If this building had been 28 feet wide instead of 30 feet could we have used 16-foot lengths instead of doing all this work connecting 8 foot and 10-foot planks?"

"That's right", he said, "We would not have to measure these boards, cut off the ends, line up the joints, use up these joining planks, use up these nails, take the time to construct these, nor put extra weight on the roof, if the building had of been designed as 28' x 42' instead of 30' x 40'.

"But if you knew that," I pondered out loud, "Why didn't you make the base that size when you poured the foundation?" "Because that is the way the proposal was sent to Canada Fund who approved the project," was the answer. We both knew who wrote the project proposal and drew up the diagrams that went with the submission. That was me!

Postscript *2019:The building is still being used for leather work, however only, Patricio, one of the original members, is still working there.*

NIGHT CREATURE

February 1994

It was in the room next door! I could hear it moving around. By the padding of the feet, and rhythm of its steps, I guessed it was about the size of a cat. Maybe it was a bush dog or an armadillo or an anteater, or maybe it was just a cat. I tried to go back to sleep. The intruder shuffled through the things I left on the porch. I tried to think if there was anything out there that I did not want scratched up.

Soon it was rooting in the bushes outside my window so to ease my mind I got up to look. For a moment I sat at the end of my bed wondering if I really wanted to find out what it was. I shone the flashlight on the door to see if it might come in and save me the trouble of getting out of bed. The noise in the shrubs continued. Out on the porch, my things were undisturbed. Movement continued in the bush.

I shone my flashlight at the place where the noise came from and spotted it at once and identified it immediately. I watched it for a while until it turned to look at me. I remained quiet and still. With the light shining on its face, I could not tell if it saw me, but it seemed to give a sign of recognition and headed towards me. I retreated! It was about ten yards away, but I was retreating. Stealing into my room, quietly but quickly, I closed the door and crawled under the sheets hoping I was not going to get a visit. There are not too many creatures I will back away from; even snakes I chase with a stick. This brute was in a class of its own. It was a skunk!

WATCHING THE HORSE DIE

February 1994

Manfred left an old horse on the farm. I found it lying in the shrubs along the pathway, between the two ponds. It tried to get up but could not. I brought it water to drink and went to report to Manfred. He wanted to know if the dying horse was near the house. When I replied no he said that I should not worry. "It will smell and attract vultures," he matter-of-factly informed me. "Can't you go have a look at it or call the vet?" I queried. He looked at me as if to say "What would I want to do that for?"

Feb 2:When I arrived home that afternoon the horse looked dead; cattle egrets were picking at it. But they were just feasting on ticks; the horse was still breathing. I bathed it in water from the pond to keep it cool.

Feb 3:When I came out in the morning half a dozen vultures took flight. I was reminded of the story "The Red Pony" where the buzzard pokes the eye out of the horse. The eye of the horse was gone. Flies buzzed around.

Feb 4:Hundreds of vultures sat around, drying their wings. They waited around the horse and in the trees. What were they waiting for? There was no access into the horse to get at the meat. I gathered that the hide is too tough to penetrate with their beaks. This story should be titled "Watching the Horse Decay". The smell was bad; I held my breath and hurried by.

Feb 5:Holes had been made in the horse's crotch and under the torso where the legs join. The smell was really bad - for about 200 yards. Birds have a poor sense of smell; it did not bother them.

Feb 6:Whup, whup, whup, went the wings of the buzzards as they ascended into the nearby trees whenever anyone approached. Only about fifty vultures were still around. The horse still had its hide, but the carcass was a shell. There was a split in its underside; the vultures must have crawled inside to "empty" it. Most sat in the trees.

Feb 7:Only half a dozen vultures remained; the smell has finally subsided but not disappeared. Vertebrae show through on the back; pelvic bones and leg bones poke through the skin; the rib cage pressed against the now thin hide. The carcass had been shifted 80 degrees and moved several feet.

<u>Feb 8</u>:No vultures were around. A couple of egrets and herons were in for slim pickings. The ribs and sternum protruded.

<u>Feb 9</u>:The leg bones had been pulled from the skeleton; flies abounded, and the smell persisted.

<u>Feb 11</u>: I moved. Someone else could live there.

<u>Feb 25</u>: Someone else did move in. There was a sign at the entrance: *"Dead Horse Ranch"*.

A GATHERING OF EGRETS

February 1994

In the early evening, the egrets gathered. Cattle egrets are a long white-necked bird that is seen as companions to livestock in the fields, mostly cows. One or two can be seen sitting on the back or near each animal throughout the "winter" months.

I used to think that the same egrets stayed with the same beasts throughout their stay. When canoeing down the Macal River at sunset, I often noticed hundreds of them clustered in the overhanging branches of the giant trees. They must proceed there to roost at night.

While walking home, one evening, just before sunset, I noticed a flock of them flying by, gathering up their brethren at their individually assigned placements. This led me to wonder how they decide which egret gets which posting?

Do they reverse in the morning; fly around to all the fields assigning egrets to cows? Do the same egrets go to the same fields and the flocking is just an escort service? Do they discuss their assignments among themselves in the evenings? It was better to wonder these things as I walked home, than to worry whether I had been robbed again.

UP THE CREEK WITHOUT A PADDLE

March 1994

These days I am living at two places. In Punta Gorda, in the south of Belize, I stay at a place that is about 45 feet from the Gulf of Honduras. I wake up to the sun rising between palm trees along the Caribbean. When in Cayo, I stay at Milli and Tina's place on the Macal River, across from Chaa Creek Resort, seven miles by road from Cayo.

Sometimes I hitchhike in or walk the Cristo Rey Road. Sometimes I crossed the river and rode the school bus with Tina's son to Succotz and then catch a bus on the Western Highway. Other times I came by dugout canoe with Mr. Green, a pleasant forty-five-minute cruise downstream. I cannot return by boat as Mr. Green heads back at noon.

Milli works in Belize City, at Crystal Auto, and has three days off every two weeks. She had asked a hired hand to meet her with the horse where the bus drops her at the Chaa Creek turnoff, but the horse was ill (not dying just ill). So he brought his dugout, with a small outboard motor, into Cayo that evening. I waited in town for a ride back by boat. It was a moonless night, and I was wondering how Andrew was going to see his way up the six miles of river. We had eight sets of rapids to pass through, but I had confidence in his navigational skills.

We made it through the first set of rapids okay; the second set I helped pole through - labourious work. If we had trouble with the second set of rapids, I was positive we would really have trouble when we reached Monkey Falls.

Sure enough, about halfway through we were at a standstill; the current pulled as much as the propeller pushed. I poled; Milli poled; after a very slow exhausting effort, we were through. I thought "good, that was the worst of it" as Monkey Falls is the toughest.

The next set of rapids gave us equal trouble - I did not recall having these difficulties when canoeing but then I remembered I would get out and pull the canoe, which is considerably lighter than a dugout. For all our poling we could not make progress. Then the propeller bottomed out. The pin had sheared. We drifted downstream.

It occurred to me, clever fellow that I thought I was, I was going to have to get out and pull. This I did. The water was surprisingly warm, and this was fine until the water was over my waist. Getting clear of the rapids was one thing. Getting back into the dugout, without tipping it, generated another fear. Andrew and I poled until we reached the next set of rapids. From my canoe trips, I knew them well.

We were up the proverbial creek without a paddle. It was slow going, and I got out several more times to pull us through. We arrived home before midnight.

I was thinking if I had not been along; Andrew would have had a tough time getting through. The next day Andrew fabricated a new shear pin and went into town during daylight, but the river was still running strong.

When he returned I asked him "How come you were able to go down and come back without any trouble during the daytime?" To which he matter of factually replied: "I didn't have your weight along." There is a metaphor for development in this story. You think your presence is helping someone out of a jam, but maybe they would not have been in this jam if you were not around in the first place.

STORYTIME

March 1994

Milli and Tina have three children between them, Jason (5), Aimee (3) and Val aged (2). When I stay at their home, I get asked the same things over and over: "Are you lighting the lamp?" "I'm lighting the lamp." "Are you lighting the lamp?" Yes, I'm lighting the lamp." Are you lighting the lamp?" And all three times it is by the same child.

"Why you lighting the lamp?"

"I want to see."

"Why you lighting the lamp huh?"

"Because I want to see, huh."

"Why you lighting the lamp huh?"

"To see- huh? To see."

"You do your laundry?"

"Yes, I'm doing my laundry."

"You do your laundry?"

"I'm doing my laundry."

"You doing your laundry?"

"Yes."

"Why you do that?"

"Because it's dirty."

"Why you do that?"

"Because I want it clean."

"Why you do that?"

"Because a man came through the window."

"Huh?"

"Is that your bike?"

"Yes."

"Is that your bike?"

"Yes, that's my bike."

"Is that your bike?"

"That's my bike, yes that's my bike, and it's my bike that it is..."

"Why you put your bike in here?"

"Because."

"Why you put your bike in here?"

"Because."

"Why you put your bike here?"

"Where should I put it?"

"Huh?"

If they cannot think of questions to ask they chant my name "Terreee, Terreee, Terreee...", and I chant back "Ayemeeeee, Ayemeeeee, Ayemeeee ..." or "Jason, Jason, Jason ..." or "Valerie, Valera, Valereeeeeeeee..."

To which they counter "Terreee, Terreee, Terreee..." for as long as I can play.

Then there is the "I wanna" variations: "I wanna a swing. Me wanna swing, I wanna swing. And when I give in I am rewarded with choruses of: "Me turn", "Again" or "Swing me like Jason got" until I cannot stand it any longer and I repeat "Tomorrow, tomorrow"

Oh, and the girls have such high-pitched ear piercing screams they can summon at will. When I try to humiliate them by imitating them, I just hurt my voice box, and they squeal in delight and then go back to murdering my eardrums. Ignoring is to no avail; they know I cannot stand it.

"I wanna story... I wanna story... I wanna story."

"After supper," I try to defer; I can read the same stories just so many times.

"Read me story", one of them persists.

"Okay" I give in "Go get the others." Besides if I wait until after supper, Val usually has fallen asleep. "Aimee and Val have assembled on the bed.

"Where's Jason?" I want to know. "He's sleepin'," claims Aimee.

"He cannot be sleeping." I exclaim ", it's not even supper yet."

"He's sleeping," affirms Aimee.

"I'll go and see," only to find Jason doing homework. "I lied," says Aimee.

"Read me story ... read me story ... read me story ... read me story ..." they persist. They like my Sea-Hag voice.

What is it about these stories that females have most of the evil roles? In Sleeping Beauty there is the Wicked Witch; in Hansel and Gretel there is a wicked step-mother and wicked woman in the candy house; in Beauty and the Beast there are wicked sisters; in Snow White another wicked witch; in Cinderella there is a wicked step-mother and wicked step-sisters, and in the Little Mermaid there is a wicked Sea-Hag.

I'd like some Robert Munch stories. I ought to be able to remember the Paper Bag Princess, but would it go over without pictures?

I drew some stick figures and told it with hand-drawn accessories. They enjoyed it, but the theme went over their heads.

One time I put off stories until after supper but Jason was still doing homework. Tina was not going to let him go until the homework was done, but Aimee had been promised a story right after supper.

If I did not read the story right away there would be high pitched yell to pay; if I read the story, without waiting for Jason, there'd be high pitched whining to listen to. I resolved the dilemma by having a coffee, thus prolonging supper.

Jason was doing his homework on the wooden bench, but the kerosene lamp on the table made it hard for him to see. I moved the lamp onto the bench between us. Jason finished his copying assignment before I finished my coffee. He got up. Like a teeter-totter, the bench went up, and the lamp came sliding towards me as I got dumped.

When the six stories had run their course I improvised: "Goldilocks and the Seven Dwarfs," "Snow White and the Three Bears," Little Red Cinderella," "The Sea-Hag Meets the Big Bad Wolf"... What would happen if Sleeping Beauty went to sleep in 1650 in her castle and everyone woke up in 1950?

Postscript: After I sent this story-letter to friends the kids received scores of books.

A TALE OF TWO VILLAGE

April 1994
SAN MIGUEL

San Miguel, one of the Maya villages, is part of the Toledo Eco Tourism Association (TEA) guesthouse program. The Chairman of TEA urged me to visit. I caught the 5:00 a.m. Z-Line bus to the Silver Creek Junction, twenty miles and one hour from Punta Gorda. I walked the six miles to San Miguel.

A teacher at San Miguel and her daughter were also exiting the bus. I helped by carrying her bags. Soon her daughter tired and rode on my shoulders. Three-year-olds get heavy after a few miles, and within half an hour of the uphill trek, the heat made itself known. As we finally sighted the village, thankfully, the little girl wanted to walk again.

Men of the village were thatching a roof for a new building. I was allowed to assist. A few days before, they had prepared hundreds of cohune palm leaves to dry them and reduce the weight from their moisture content. These 20-foot palm leaves were split down the middle lengthwise and passed up in bunches of three. I was invited onto the roof to see how it was done. Next, I passed palm leaves up to the men doing the thatching.

Specific sides of the palm leaves had to be passed up at certain times. They were placed so that the folded side pointed down and were alternated from left side to right side to channel the rain. I picked up a bit of *Kechi*: *wan* means "wait"; *sa kal* means "enough". They coined a new phrase when they wanted thatch passed up. "C'mon Terry". They said that even when I was not present.

Dirty from passing the dusty palm leaves, I took a swim in the Rio Grande River. Springing from a large cave, the river begins a mile upstream. One of the attractions of the village is the hike to the cave with a canoe ride downstream. Where I swam, it was deep, cool and blue. After lunch with a Maya family, I met with the craft group, to discuss a funding proposal.

In the evening I was invited to attend a church service with songs in Spanish and English. When I entered, I noticed one side had twice as many people as the other side. Going against my normal instinct to sit with the minority I sat on the left with the "crowd" thinking they knew something the rest did not.

People, particularly from my side, kept looking back at me. I thought that it could not be that unusual for a visitor to attend service. The Chairman and his spouse arrived together, but she sat on the left and he on the right. That struck me as unusual. Then a hypothesis took shape. Looking carefully at the composition of the two sides I realized there were all men on his side and all women (except for one) on the other side. I changed sides; the funny looks disappeared.

LAGUNA VILLAGE

A week later I caught the 5:00 a.m. Z-Line bus to the Laguna Village junction. It was only a three-mile walk to the village. My primary target was to see the rock art in one of the caves (I was attending the International Rock Art Congress in June). My guide tried to dissuade me, as there was no path. After I persisted, he asked me to put on long pants, as we would have to break a new trail. When I informed him I did not have any he just shook his head and fetched his machete.

The first few miles were along a well-worn trail. The guide pointed out various plants we did not have in Cayo District, including the jippa jappa plant, which is used to make baskets. He inquired about the ancient Maya: "Where did they build? Why did their civilization decline? Did they go to school?" - All good examination questions until "Where did the first Maya come from?"

After several miles, we had to head into the thick brush. He cut a narrow path suitable for his 5' frame but not my 6' 2" (220lb) bulk. The going was rough. He often stopped to ask if I wanted to rest, to which I replied "No". After a long time, we arrived at the cave, and I viewed the Maya hieroglyphs on the cave walls. They were mostly covered by slow dripping sediment along the surface.

Then the guide wanted to know how long humans had been present in Central America; where they came from; how long humans lived on Earth; how old was the Earth; how big a number was four billion; and what did the Bible say about all that? He was either very inquisitive, or it was because I was paying him by the hour. We continued to the other cave - where he normally took visitors.

We took a ladder up to the first entrance, about two stories tall to find room size cavities. After about twenty-five yards up-hill there was a "Y" junction. To the left was a tunnel with a crawlspace that led to a large chamber. To the right was a long corridor that kept going uphill. It then doubled back so that we came to another large opening above the first entrance way, about thirty feet above the first. It provided a tremendous view to the east.

Further back was a narrow, uphill corridor to another ten-foot high opening, this time facing west. Quite impressive! After lunch, we headed four miles out to the Caliente Lagoon. It was level hiking, and after we passed the milpa fields, the vegetation changed sharply to trees and brushes typical of a flood plain. At one point we could hear and see a fire used to clear land for further agriculture use.

Once we reached the river, there was an abundance of birds. There was a pygmy kingfisher and the much larger ring kingfisher.

The lagoon area itself seemed like something out of Africa - a broad low-lying grass veldt with a body of water in its center, with trees on the far side and Toledo Ridge Mountains in the background - beautiful. Orchids grew on many trees. The lagoon was full of cormorants and different types of egrets. There was one large stork. The hike was outstanding, but for some reason, I felt really tired. Then I realized I had been walking from sunup to sundown - about thirty-two kilometers in all.

BELIZE CITY IS A DANGEROUS PLACE

April 29, 1994

There were 212 murders in Belize in 1992. Most were in Belize City, population 45,000. Calgary, Canada, had 8 murders that same year. If Calgary had Belize City's murder rate, it would have had 2,000 murders! Right up there with killings is car accidents. There were deaths every week. I had never seen a dead person unless one counts the bones of Don Fabio. I had been at a meeting in Belize City and did not feel up to the bus ride back to Cayo, so I stayed with friends who lived along the northern highway. At 2 AM a crash shook the trailer I was staying in. I ran out and saw a two-door vehicle with its headlights on in the boatyard next door.

Neighbours peered from screen doors. I shouted for someone to call the police and asked the guy standing by the driver's door if anyone was hurt. He replied that one was dead. I hollered for someone to call an ambulance. I was not confident the guy would know an unconscious person from a dead one. Noting glass around the wreckage, I put on shoes. The person on the driver's side had his face pushed in, full of blood. There was a large hole in his head; no blood was coming out. The man was dead.

There was now a fellow standing by the passenger's side, who at first I took to be a passer-by. He had been in the vehicle also. While pointing to the dead man, he said things like: "We charter him" (meaning hired him to taxi them).

The car lights had been turned off. The "passenger" then asked me to shine my flashlight on the floor of the vehicle to help him find his money and wallet. It was then that I noticed the dead man's feet were on the passenger's side - a funny place to be if he had been driving - but then the car had taken a hell of a wallop. The passenger side roof was caved in, and the passenger door was crumpled flat.

I heard a moan from the backseat and realized there had been a third occupant. This guy looked as bad as the dead man. I had brought a towel with me and positioned it to his head to soak up the blood. He put it under his head and lay down. After ascertaining that the police and ambulance were on their way, I learned that the victims were from the Belize Defense Force (BDF).

There were beer bottles in the car - they had a night off and were returning to their base.

The poor guy in the back seat kept jerking up and crying out. I tried to calm him by saying, "Lie down, you'll be okay, just lie still". "Alright baas," he replied, "alright baas," and then he moaned out that he hurt. He looked plenty hurt. The other guy, outside, kept wanting to find his wallet.

Although sympathetic to his being in shock, I finally told him his wallet could wait as we had to concentrate on helping the injured man trapped in the back seat. Late nightwalkers and neighbours had gathered around.

When the ambulance arrived one of the attendants stated that they could not get the guy out of the back seat without the fire department. The driver wanted to perfect his parking and kept maneuvering the ambulance. As we had already waited thirty minutes, we tried to remove the fellow from the back seat.

There was a space between the compressed roof and what was left of the door. After assuring the ambulance driver that he had positioned the rescue vehicle just fine, we tried to extract the injured man. Each of us took hold of a thigh and tried to pull him feet first from the back seat. I was as much to blame as anyone for this useless approach. The fellow's right shoulder was out of its socket, and we could not use his arm to move him.

It would have been easier to pull him out torso first, but now he had slid to the floor, and we could not get him back on the seat. The back of the car was pressed into a boat; we could not access him that way. A passerby asked what the problem was. I explained how we were having trouble moving the injured man. "Why don't I just get in there and lift him up?"

Without any hesitation he climbed through the small space without touching the victim, positioned himself behind him, and lifted him up enough for two of us to get him onto the stretcher. The ambulance guys closed the rear doors. I asked if they could take the other guy who had been in the crash. They said "sure" but drove off without him. The hero who had climbed into the car was an off duty medic. To his credit, he checked for a pulse on the dead man.

The police arrived but decided they had to get their sergeant and a photographer. I asked if they might take the other guy, who was still looking for his wallet, to the hospital. They did, but with that casualty's resistance.

He did not seem too bad off. The blood on him looked more like it came from somebody else. With the ambulance and police gone, we bystanders were left with the wreck and the dead man. One person commented it was funny how the one guy kept saying that the dead man was the driver (when nobody was asking). I got to thinking, who would think to turn off the lights but the driver? A closer look at the destroyed car revealed that the passenger side took the impact.

It was hard to see how anyone could escape injury if they had been on that side of the car, where the windows, doors, and roof were demolished. Whereas on the driver's side, the door was stuck but still intact. The driver's window was down. Someone could have climbed out.

What about the guy's wallet - how could he lose his wallet and if the wallet was missing, how did he know there was money out of it? Besides the "passenger," the only thing pointing to the dead man being the driver was that his large body was leaning over into the driver's seat of the small car.

A reconstruction of the event began to be played out by those still on the scene. There was an all-night gas station down the road; the driver, while taking money out of his wallet for gas, was not watching the road and clipped the side of a parked car. This spun the car 180 degrees, before smashing through a concrete wall, destroying some display boats on the other side. The driver had had the steering wheel to hang on to. He climbed out as I arrived on the scene and replied to my query from the yard next door that one man was dead. He then turned off the lights and moved to the passenger side. This scenario was suggested to the police photographer.

Postscript: *The guy injured in the back seat was taken to hospital in Merida, Mexico. A brother of the dead man conducted his own investigation. He did not believe that his brother had been the driver. He was waiting to talk to the patient taken to Merida. Several years later there was a court case against the "driver".*

FURTHER UP THE CREEK

May 1994

After being stuck in the river a few times, Andrew opted for a flat-bottomed boat and larger engine. I could hear him doing test runs. That motor really hummed. We were going to the Negroman orchards at Tipu to pick oranges.

Tipu was three miles away. The first half-mile was like riding a powerboat through the everglades - smooth. That was before we reached the rapids. It was so shallow we had to walk the boat barefoot. There were rapids about every hundred yards - hardly worth having an engine never mind a more powerful one. By the time we reached the orchards my feet were sore and toes bruised from the jagged rocks, I encountered while pulling the boat. We had to walk half a mile to reach the orange trees.

There were no oranges ready to bag which meant picking our own. Oranges sold for $9 BZ a sack. Andrew located a large, older tree. He climbed up and shook oranges loose. The older the tree, the smaller the oranges; these were about baseball size. Oranges on two-year-old trees are about softball size. The older the tree, the juicier the oranges.

The bigger oranges take more space but yield less liquid. We filled a sack with 300 small oranges. From one tree we filled three bags, and part of a fourth - over 1,000 oranges from one tree and more oranges hanging on the branches of other trees.

We targeted the smaller trees (bigger oranges) for a few more sacks then waited for the tractor to carry them down to the river. Throughout the morning I had been eating oranges, but I could not quench my thirst. I was some thirsty! I hauled the oranges to the boat. It was thirsty work under a noon sun. Surrounded by oranges I had my fill. What I craved was in the fridge back at the farm.

We still had to get out of the boat for the rapids. We did not have to push and pull. At the farm, I carried oranges up the hill to the storage room and then headed to the fridge. It was there - a Coca Cola - it refreshes best!

I SAW AN ARMADILLO TODAY

As I walked the trail home to the farm, I had been seeing an armadillo about twice a week. First I heard it rooting in the trees, perhaps as close as twenty feet. The vegetation was not thick, and I could spot it seeking out insects or whatever it eats. It did not pay any attention to me; it just went about its business. I watched for a while thinking how great it was to witness this creature in a natural habitat.

Then I encountered a baby armadillo sitting in the middle of the trail. It was about the size of a kitten and reminded me of a kitten except it looked like a really cute armadillo.

It was only a pace away. I could have bent over and scooped it up in my hand, but I just observed. It seemed to be preening. After a while, it noticed me and looked up. The armadillo did not know what to make of the huge beast sharing the pathway. I wanted to pick it up and bring it to the farm to make it a pet.

As we gazed at each other, I visualized it running around with the puppies, the cats, and the chickens. I could hear Tina saying "Jason, please give water to the dogs; Jason go feed the cats; Jason feed the armadillo."

No, not another animal to look after when it could look after itself. Just then the little guy scampered up the embankment and turned to look at me again. It paused with a query on its face as if to ask "Would you really turn me into a pet?"

Then it shuffled off another ten feet or so before looking back again. I decided I wanted to see big armadillos rooting around in the bush I had better let little armadillos grow up to be big armadillos.

MORE OR LESS

At school, Aimee and Okie were learning about more and less. It was part of a module at school for teaching about quantities: when something is more when something is less.

You do not have to teach kids about more or less. Just give them some popcorn, or pour out sugar frosted cornflakes and there are instant shouts of "s/he got more." Or give one a swing and its "she got more" or "he got a longer turn". But tell them it is their turn to clean up and the concept of equality disappears. They understand - more or less.

DELIVERING BANANAS

October 1994

On the way back from the annual SPEAR Conference (Society for Promoting Education and Research) held in Belize City, I disembarked the bus at the Belmopan turn-off. I hoped to save the half-hour the bus takes to swing into the Capital. Right away I caught a ride with a van delivering bananas.

This fellow delivered bananas every Monday to shops along the Western Highway. I became his assistant for these deliveries. He would buy green bananas the week before, for $600 a truckload, which he puts into boxes after ripening them in a specially heated room. He gets two vanloads per truckload.

He had about a hundred boxes in his van, which sold for ten dollars a box. That meant he paid about five cents each per banana and sold them for ten cents each. On the street, in San Ignacio, they sold two for a quarter or seven for a dollar (yes it is cheaper to buy them two at a time.)

One of the stops was at the home of the President of the Teakettle Survival Group. Another interesting coincidence was that this gentleman was married to a sister of the Treasurer of the Branch Mouth Women's Group. He and his wife put their ten children through school (two were already in college).

Bananas were cheap because of how immigrant labour is exploited. This was one of the topics covered at the SPEAR Conference. There was a discussion that Belize did not have sufficient population to support a banana industry. Accusations were made that not only was the labour force poorly paid but also they were brutalized if they complained. Here, as in other banana-producing nations, was an industry that was monopolized and subsidized and utilized a poorly paid and poorly cared for workforce.

From a developmental point of view, it did not make sense. We spend money to get small producers in business. These same countries buy from multi-nationals that abuse workers. Why can't the development process buy bananas (something we want), so that it directly helps the small producer?

HIKE AND BIKE (TOSS) FOR THE RAINFOREST

October 30, 1994

Accion Selva is a Cayo based organization involved with putting on the <u>Hike and Bike for the Rainforest</u>, a competition with funds going to the Belize Zoo and Audubon Society. There were nine corporate sponsors. There were twenty-five entrants who had each paid $200 to compete in the running and bicycle race. That fee eliminated me, so I volunteered.

Saturday was the day of the eight-kilometer run being held at Chaa Creek Resort. It was across the river and down from where I was staying. In the morning I paddled the dugout to the Chaa Creek landing, hooked up to one of the Chaa Creek canoes towed it back, returned the dugout and then back to Chaa Creek with their canoe. A group of us volunteers were herded into the back of a pick-up and dropped along the route to guide runners.

One of the competitors was also along. He was not going to do the hike, just the bike part. When asked about that he replied that he was not crazy. He was "big O," obnoxious. It takes one to know one, I guess. The first drop was at the halfway point, furthest from the starting point - no one was anxious to stay there. I took it.

I was glad to get away from Mr. Obnoxious. It turned out to be one of three locations where water was provided for the contestants. A little while later, a fellow was dropped off to distribute the water. Now I had company and maybe more important - water. The first runner was along almost immediately.

They must have started just after we left in the pick-up. The first runner was so far out in front, that at first I thought they were giving staggered starts. Then the others followed in twos and threes.

"You are halfway," I shouted encouragingly, "just up this little hill". I pointed to my right. Except for one fellow (who ran up) they would take the water from Lenny, turn the corner, see the hill and walk up. I guess it wasn't so little. The last participant had a race official following her, so we knew that they were all through. After Lenny and I picked up the emptied water bottles along the pathway, I headed back along the trail to pick up other items dropped along the way.

After the truck picked up the empty bottles, the other fellow caught up with me and we backtracked together. He had wanted to enter but had a bad knee from jumping out of a truck with a loaded pack when he was in the army.

Along the way, we found plastic water bottles and tops. The trail went up and down hills through the forest. It was well marked; every once and a while there was a horse trail sign where the numbers kept going down.

We had started at #29. It seemed to take a long time to reach #10, but then the next sign was #19. I did not start to be concerned until we saw #15 for the 3rd time.

NOT GETTING ANYWHERE

The annoying thing about walking in circles is not the wasted time, but the realizing you are walking in circles. There were a couple of places where the trail looped around on itself after it crossed a bulldozed track. The other fellow thought he knew the way back. After the fourth time around I figured the bulldozer had to have come in from a road. As we were at the end of the track, all we had to do was follow it out. That worked. We were longer getting back than it took to do the whole course.

HALLOWE'EN III

After a lunch, hosted by Chaa Creek, I caught a ride into San Ignacio to attend some of the evening ceremonies. It was also the night of the annual Hallowe'en party at Caesar's. The costumes were not as good as previous years, and more people came without "get-ups". Milli and Tina were on my case about entering the costume contest.

I suppose that since they lent me a sarong, bikini top, and wig, I should have entered. What is the point of being judged if you do not want to win? What if someone recognized me? Anyway, first prize was an overnight stay at Black Rock Resort (further up the river from us). It would be like living in Banff and winning a pass to ski at Lake Louise.

When I went to the bar to get a drink, there was a long line up. I waited my turn and asked the bartender for a Coca-Cola. Just then a flaming match arced over the bar top into the bottles of Caribbean Rum. I looked to see who the "A___ole" was, and there was Mr. O. He said to the bartender, "I just wanted to get your attention - could you put my bag behind the bar?" I kept my mouth shut. I thought, "This guy is in the race tomorrow, and maybe he is going to depend on me along the way". It will be a difficult enough course for him. The bartender took a while to get re-orientated, but he served my coke.

UP AND DOWN THE HILLS

The fifty-kilometer bike race started from the Piache Hotel. Mr. O. was there in the back of the pack wearing a T-shirt with rats on it. The first part of the race was up the Cahal Pech hill; I did not even like walking up that sucker. Some contestants were walking their bikes already. I dropped volunteers off at Clarissa Falls and Calla Creek.

Just before Clarissa Falls was a hill that commanded a fairly good view of part of the course. From San Ignacio to this hill the competitors would be following horse trails. I left two of the race officials there and continued onto Calla Creek. I noticed there was a "Y" branching of the road that might confuse the cyclists. Going back I used a can of spray paint to mark the gravel with an arrow. The first cyclist on the trail had complained about lack of markings.

A SWINGING BRIDGE

At Calla Creek, a narrow swing bridge crossed the Mopan River. On the other side were bottles of water for contestants. There was also an army helicopter. I think it was being used to help take photographs for news magazines and television coverage. As I walked across the planked footbridge, I noticed several boards missing.

On the other side a race official, with a radio, advised me that one person had already dropped out of the race (giving up on Cahal Pech hill). The bridge swung very easily. I left my wallet in the truck in case I had to dive in to help anyone falling into the river. The first problem was getting them to the bridge.

Just before the bridge, it was difficult to determine which way to go. I spray painted arrows on the gravel road and stood there shouted "just over the bridge," and signaled with a circular motion of my arm and said, "You are half way". Then I added, "take it easy over the bridge," and ran up to the bridge to see if they were crossing okay. I would do the same for Mr. O! Some competitors walked their bikes, some rode slowly. One had a bit of trouble and almost went over the edge.

Checkers on each side wrote down numbers of each contestant and their crossing time. If they were missing their numbers, the race officials usually knew their names. One cyclist just went charging across the bridge at full speed. If there had been a prize for the fastest across the bridge, he would have received it. Villagers, officials and even the camera crew all gave him a big cheer. I looked on the checkers sheet to see what number was written - he did not have a number - instead, she had written "Animal".

The helicopter lifted off to take aerial shots of cyclists crossing the bridge. At one point it lowered to a few feet above the river so that the photographer could get shots directly level with the bridge. Blades of the chopper seemed to cut the grass on both embankments. The river water swirled beneath it, but the chopper held steady.

It then took off to film the riders at a point further along the course. The next section was probably the roughest part: a crude trail paralleling the river into Bullet Tree Falls. From Bullet Tree Falls they would get to take a five-kilometer road back to San Ignacio.

Twenty-four participants crossed the bridge but no Mr. O. It started to rain. Half a dozen of the volunteers piled onto the truck, and we headed back to town. At San Ignacio, we had to wait while the police waved one of the riders through. Eight of them arrived before us.

There were prizes for best cycling time, best running time, best combination; and first, second and third prize awards for winners in different age categories. For instance, there was only one person in the 40-44 age group. By finishing I could have placed second. There was still the bike toss.

TWIRL AROUND TWICE THEN LET GO

When I arrived at the celebration tent, I entered a tournament to see who could throw a bicycle the furthest. I had experience with this sort of thing.

In October of 1991, a friend and I had gone on a bike trip to Mount Assiniboine. We had made a pre-dawn start from Chain Lakes. We arrived at Mount Assiniboine Lake as the last visitors were leaving.

They laughed at my three-speed, but I made it without the helicopter lift, unlike most of them who had their 18-speeds brought in by helicopter. Normally, one stayed overnight. My friend did not know that nor did I. He was annoyed with me, as this was a terrific spot - perhaps the most beautiful place in the world. It was more picturesque than Lake Atitlan, Guatemala, which is billed as the most beautiful lake in the world.

Now that we had just reached the destination we had to turn around and head back. Not satisfied with the splendid scenery along the way, and the beautiful scenery at the top, we took a different route back. We returned along a different set of trails.

If I had not taken us down a dead-end pathway, we might have returned before midnight. That way down was not meant for bicycles. On one part a high growing root snagged my pedal, and I flew over the handlebars.

After three such flips, I was resigned to pushing my bike down the hill. But the path was not wide enough. The clasps of my hiking boots kept catching on each other causing me to fall forward with the bike falling on top of me. After one of these exchanges, I flung my pack and then the bike into the bush.

My friend, who was bringing up the rear, laughed uproariously. He had watched me stumble, the bike drop on top of me, my pack being taken off and flying away, the bike being picked up and joining the pack. I went into the bike toss with experience.

First prize was a Hike and Bike T-shirt. The first person to throw achieved twelve yards. The second thrower got eleven. My toss was nineteen yards. After each toss, the bike became a bit lighter. A pedal fell off, a spoke popped out. I did not wait around; I had my fling. Besides I had a T-shirt from being a volunteer.

Lady El

Back at the farm the first to greet me, as always and with enthusiasm, was Beauty one of the guard dogs. She did not bark at my arrival. The other guard dog, Muffin barked at anything and everything. The puppies, Baby Bear and Princess, guard dogs in training, came wagging their tails. The chickens did not acknowledge me; the cats ran away.

Tied up, along with the dogs, was a two-month-old pig. She used to be kept in her pen all the time. I implored her captors to let her have some freedom. She used to lean against the gate of her pen every day wanting to get out. I thought if she had a name it would humanize her and maybe exempt her from the inevitable. After I christened her Lady Elegant, she at least was allowed to roam the yard. She seemed to think she was a dog as well as she waited for her pat on the head like the others.

IMPENDING DOOM

November 8, 1994

I could hear it a thousand yards away. At first, it was just a distant rumble, hardly noticeable, hardly interrupting my thoughts as I made my way down the path. At about 800 yards the volume picked up enough to make me realize "It is on its way!" At 600 yards, there was no doubt. The noise became louder and louder, sufficient enough for me to turn around and look. Dark, ominous clouds were looming. It was obvious what I was in for, and there was nothing I could do about it. I had too far to go.

Dark and ominous just does not cover it. With 400 yards before impact, I was not concerned with what to call it but with what to do. My impulse was to run; it would not serve any purpose. I tried to rationalize I was going to be home soon; I was going to do my laundry anyway. At 200 yards the noise was an encompassing thunder. The rainforest was succumbing to the onslaught. I held my pace. At 100 yards the sense of impending doom was overwhelming. It would not do any good, but I did what my reflexes were crying out for - I ran. That gained me a few seconds reprieve. Then the downpour, or rather, waterfall enveloped me. It was an instant soaking accompanied by a deafening clamour - a precipitous greeting.

ILL CONCEIVED, POORLY THOUGHT OUT

November 19, 1994

Ill-conceived and poorly thought out! I had this brainstorm to take Aimee, Milli's daughter, to Belize City to visit Milli at work. Wouldn't it be a nice surprise, I thought? Aimee would like that, and Milli would like that...

Yes, Aimee liked that idea. First I checked with Tina. Okay, that is settled. "Pack a change of clothes," I advised, and I went to my room to get a book. When I came out Jason was crying; Okie was crying. Now, this did not figure (to me). Why should they be crying about Aimee going to town? But that was why they were crying - because Aimee was getting to go to town and they were not. They felt slighted, left out and traumatized.

First I tried a rational approach. I picked up Okie and explained that this was a chance for Aimee to be with her mother. "Don't you want to be with your Mom here on the farm?" I inquired. I set her down thinking that I triumphed with logic. She was still sobbing, so Tina hugged her and asked, "Don't you want to stay here with me?" A tearful Okie shook her head.

Let me try and work on Jason. "Jason, what are you crying for? Don't you want Aimee to have a chance to be with Milli? Sobs and bawling were the reply. Aimee was happily packing. Tina was concerned. Both of her children were crying, and I was perturbed because I was the perpetrator.

Okie had a reasonable gripe: Jason had gone to Belize City with Andrew just last weekend. She figured her and Aimee should be able to go this time. "That's right," I tried to reason with Jason, "you got to go last weekend, and Okie and Aimee did not cry then." That gained no points.

"Alright how about if nobody gets to go?" I challenged Jason. Tina liked that idea, and she tried to soothe Okie and Aimee with the promise that Milli would be home soon anyway. I took a coward's exit and went to the bathroom. When I came out, Jason had stopped crying, but now Aimee was crying and so was Okie. This was a no-win situation I had created. I was solely responsible for upsetting everybody. I felt as if had betrayed Aimee, having offered her a trip to the "big city" to be with her mom, and Okie looked so sad with tears streaming down her face, from an opportunity snatched away from her. I took Tina aside and said that I could handle two kids.

Anyway, Jason had his trip; he should appreciate fairness and a chance to be alone with his mom. Tina relented.

She packed for the girls. I took Jason aside and talked to him man-to-man about his being the only one to help out on the farm and to keep his mom company. He stopped crying. He looked at me, his face formed a pout, and he stomped off to his mom. Tina said firmly "you are not going." Jason wailed and went to his room, and started packing his bag.

Jason raced ahead and was the first to climb into the 4WD. Tina hauled him out. I strapped both Aimee and Okie into the front seat (I did not want Ms. Mischief behind me). Jason climbed in behind the driver's seat. I hauled him out. He was not happy. Aimee and Okie were happy. As the 4WD engaged, I shouted to Tina "Have a nice day!"

The kids behaved well during the two-hour ride to Belize City. When I arrived at Crystal Auto, I expected Milli would be all smiles. She smiled at the girls all right, but all I received was a stare communicating "Buster you are going to be a baby-sitter for the day".

I can handle two kids. No problemas! We started watching the cartoon network. I could handle an afternoon of cartoons. I was curious to see if children who had not been exposed to television would be engrossed. I left to fetch lunch with instructions to "Stay, in, the, room." I returned to find Milli surrounded at her workplace. I retrieved them with the lure of soft drinks. Fascination with the cartoons wore off whenever a commercial came on. Okie discovered electric lighting. The lamp went on; the lamp went off on… After thirty ons and offs, I wondered how long this could go on? Aimee came to the rescue and said: "Okie you'll run down the battery." Not technically correct but Okie ceased the activity.

After the bathroom visit, Aimee discovered that by throwing clothes in front of the floor fan they float around the room. It seemed like a pretty harmless activity (there is a protective screen). I turned the floor fan so that it pointed straight up. Soon Aimee and Okie emptied all their carefully folded clothes out of their bags. They "filled" them with air and launched them skyward.

While they were preoccupied, I switched to the Discovery Channel. Okie wanted to play with the TV channel changer. I showed her how to change channels back and forth. I figured we would watch the Discovery Channel half the time, and view cartoons half the time.

That worked until Okie located the "mute" button. I explained that we could not hear anymore. She understood just fine. It was mute on, mute off, mute on, mute off.

It was not long before she experimented with the other buttons: channel up, channel down, channel up, channel up, channel up, channel up. That went on and on with Aimee and Okie taking turns until … bathroom time.

They had been learning numbers at school. I brought out the calculator. "See this has the same numbers as the channel changer" as I implied that playing with it would have the same effect as playing with the remote. "See, what number is that?" as I pointed to the number 8, one of Aimee's favourite numbers. "Now push it to see what happens - you get an 8 up here see?" That had her attention for all of eight seconds.

Okie kept disappearing around the corner. I looked to see what the attraction was. She was intrigued with a small bike near the wall. They could not reach the pedals, but they were delighted taking turns being pushed around. As I had neglected to bring toothbrushes along, I was delegated to remedy the oversight so I took the kids to the store, hoping the walk would wear them down. They fell asleep after brushing their teeth.

A great time was had watching cartoons, going to the bathroom, playing with the fan, going to the bathroom, playing with the light, going to the bathroom, playing with the channel changer, playing with the calculator, going to the bathroom, "riding" the bicycle… Jason had a good time on the farm. Ill-conceived, poorly thought out, but well executed.

Politics and Crime

Following World War II, the economy languished. Britain's devaluation of the British Honduras dollar worsened economic conditions. This led to the formation of political parties.

The party system in Belize has been dominated by the leftish People's United Party (PUP) and the rightish United Democratic Party (UDP). The first election held in 1954, was won by the PUP. George Price became head of government from 1961 to 1984.

Although Britain granted British Honduras self-government in 1964 it was not renamed Belize until 1973. Even though Belize achieved independence on September 21, 1981, Guatemala refused to recognize the new nation because of a territorial dispute. British troops remained in Belize to discourage Guatemalan incursions.

In 1984, the PUP were defeated by the United Democratic Party (UDP). Manuel Esquivel replaced George Price as prime minister. The PUP returned to power in 1989. In 1990 Britain announced it was ending military support in Belize. British soldiers left in 1994 but provided military training for the newly formed Belize Defence Force.

The UDP regained power in 1993; Esquivel became prime minister for a second time. Esquivel, perhaps unwisely, announced the suspension of a pact reached with Guatemala during Price's tenure. He claimed that Price had made too many concessions to gain Guatemalan recognition.

In 1998 the PUP won a landslide victory, with PUP leader Said Musa becoming prime minister, and again in the 2003 elections. Musa pledged to improve conditions in underdeveloped parts of the country. Tax increases led to discontent with the PUP government. The UDP won in 2008, with Dean Barrow becoming prime minister. The UDP were re-elected in 2012 and again in 2015. The next election is scheduled for November 2020.

Each party has its own newspaper. They are quick to point out corruption of the opposing party. For instance in the Amandala August 17, 2019 edition, readers are reminded of the loss of a new swing bridge for Belize City, funded by mainland China. The whole bridge went to Jamaica, when the PUP came to power and established diplomatic relations with Taiwan, in 1989.

The Taiwanese have supported Belize with hundreds of millions of dollars to help with budget shortfalls, [road building,] providing youth scholarships and transferring agricultural technology. Whereas the article stated that mainland Chinese [immigrants] have taken over grocery stores, fast food outlets and gambling. With their families operating such businesses there have been less employment opportunities for Belizean youth.

The Amandala article went on to mention that in the 1989 election PUP had accused the UDP of registering mainland China passport buyers as voters. "But Taiwan has gotten way too chummy with the UDP, and somewhere along the way, Taiwan's assistance started helping UDP more than it helped the people" was a claim in the Belize Times.

Although the tourist and construction sectors have strengthened, poor controlling of spending has brought the exchange rate under pressure. Infrastructure is a major economic development challenge. Despite promises to the contrary Belize has the region's most expensive electricity.

Belize Electricity Limited (BEL) is largely owned by Fortis, a Canadian company. Fortis took over its management in 1999, to address financial problems from being locally managed. Fortis also owns and operates three hydroelectric generating facilities on the Macal River. These have been considered environmental calamities.

Fortis also has ecological issues with its Muskrat Falls operation in Labrador where it is accused of methylmercury poisoning of indigenous food supply downstream. The Canadian government has invested $9.2 billion in the project without the prior and informed consent of all peoples affected. [Change.org August 26, 2019]

Indigenous Land Claims
Belize has done little to recognize Maya land rights and does not even have a land registry for them.

Guatemala's Territorial Claim
Guatemala has claimed ownership of all or part of Belizean territory. This claim is reflected in maps by the Guatemala's government, showing Belize as a province of Guatemala.

As of 2019, the border dispute remains unresolved. Guatemala's claim to Belizean territory rests, in part, on the Anglo-Guatemalan Treaty of 1859, where the British offered to build a road between Belize and Guatemala. Guatemala became more interested in American funding for banana plantations and seemingly forgot about the road, until Belize achieved independence.

In April 2018, Guatemala held a referendum to determine if the country should take its territorial claim on Belize to the International Court of Justice (ICJ). One quarter, of 7.5 million Guatemalans eligible to vote, turned out, with 96% voting "Yes". A similar referendum held in Belize, in May 2019, had two thirds of 150,000 eligible voters turn out. Only 55% favoured sending the matter to the ICJ. It may be more than two years before the ICJ makes a ruling.

Just the southern portion of Belize is deemed at risk of ceding to Guatemala.

Crime

Belize's geography has made the country's coastline and jungle attractive to drug smugglers. In the seventies it had a reputation for growing marijuana (*Belizean Breeze*) where planes would land on highways for pick up. Spraying against killer bees invading from South America apparently subdued the marijuana industry. Illicit drugs have been suspected of being included with logging shipments from Guatemala. Belize is used as a gateway to Mexico. With the Belize currency tied to the U.S. dollar, drug cartels use banks to launder money, which is facilitated by allowing foreigners to do offshore banking.

Belize has high rates of violent crime. The majority of violence in Belize stems from gang activity, which includes trafficking of drugs and people. The Belize Police Department have implemented protective measures to address the high number of crimes, such as: gaining more resources, adding more patrols, creating preventive programs and setting up a Crime Information Hotline.

In 2018, there were 143 murders in Belize. Two-thirds of the killings occurred in or near Belize City. This equates to a homicide rate of 36 murders per 100,000 inhabitants. Although lower than the 100+ per hundred thousand of 1992 (as noted in the story "Belize City is a Dangerous Place") it is still one of the highest in the world. The rates in Honduras and El Salvador are sorrowfully even higher.

A government established truce among many major gangs, led to a lowering the murder rate. However there are still many reported cases of rape, robberies, burglaries and theft. While in Belize my home was robbed at two different locations (as noted in the stories "Robbed", "The Woman with the Golden Bag" and "On the Trail of the Golden Bag"). Six of the people I worked with or had as friends had been murdered during my stay there or in the years after ("The Bones of Don Fabio" tells of one death). A fellow Cooperant was run over by a drunk driver (a form of homicide).

1995 – The Post Classic Period

Girls Are Smarter Than Boys

THE WOMAN WITH THE GOLDEN BAG

January 9, 1995

In the mornings Batty busses travel west while Novelo's buses travel east. The yellow bus is an express that leaves Belize City at 6:30 a.m. It reaches Belmopan fifty miles away at 7:30 a.m., in time to catch the white bus, which leaves at 6:00 a.m. The white bus reaches San Ignacio by 9:00 a.m.

A year ago this month I was robbed. Everything was taken but my books and clothes, $4,000 worth of furniture, household items, personal effects, and souvenirs. Nothing had been seen of the hundred plus items that had been stolen - until this morning. While sitting on the white bus, two miles from San Ignacio, I looked out the window. From my rear sea, I noticed a man and a woman standing on the other side of the road. They were standing on either side of a golden yellow suit bag. The man crossed the street to get on the bus; he brought two bags of oranges. I thought he was going to help the woman with the golden bag. It looked like the suit bag I had a year ago.

The bus pulled away. The man had not helped her on. Looking out the back window I had a better view of the bag. It was missing the top clasp, the same as my bag! As the bus proceeded to Cayo, I realized that the intersection road she was standing at was the road to Pilgrim Valley. This was the road that leads to the trail backing onto the place where I used to live - the place I was robbed from. I watched where he got off so I could ask about the woman with the golden bag.

He got off at San Ignacio. His name was Daddario. He claimed not to know who the woman was. "Come on now, you were standing right beside her," I challenged, "Your bags were touching her {my?} bag." He insisted he did not know her. At the CCDC office, I phoned the police. The line was busy. I tried to borrow a CCDC vehicle. The truck and one bike were being used, and the other bike was broken down. I walked to the police station. They did not have a vehicle at hand, but an orange Novelo's bus was just leaving. Three officers and I piled on - if she was going east, this could be her bus.

Back at the Pilgrim Valley junction there was no woman and no golden bag. Up the road, we saw several people walking with bags but too far away to distinguish colours. A police vehicle with two more officers joined the chase. We four climbed in and went along the road. A yellow Batty Bus headed west. "Yes," the man said, in Spanish, he had seen the woman "she got on a white bus going west". We followed.

There were two officers in front; one on either side; or one behind me in pursuit of a woman with a gold suit bag, on a white bus. Maybe it was a different white bus; I doubted it. We were being misinformed.

THIS BUS, THAT BUS, WHICH BUS?

I tried to explain that she could not be on the white bus, as I had been on the white bus. The next bus, which passed when we were at the junction, was a yellow bus (the former express). It was to no avail.

We went all the way to the Guatemala border before catching up with the white bus. It was the same white bus I had been on, with the same driver and the same conductor - no woman with a golden bag. We headed back to Cayo. That was the end of the pursuit. There was no gas remaining in the police vehicle.

INVESTIGATING ON MY OWN

Back at the office, I continued with the one-on-one lessons with the Field Officer I am training in Economic Analysis. I remembered a Mennonite man was standing up the road away from where the woman was standing. I recalled seeing him at Manfred's place. After the economic analysis training session was over, I left a message for Manfred at Macal Dairies. He came by shortly after and told me that the Mennonite lived one mile up the Pilgrim Valley road.

At about 3:00 p.m. the Yamaha 550 became available. The Coordinator and I went along the Pilgrim Valley road. The Coordinator helped with the Spanish. We found the Mennonite gentleman at home. Yes, he remembered the woman. She was the girlfriend of one of the fellows living further up the road. They had taken a blue Shaw bus going to Belmopan. The fellow was on his way back to San Pedro to work. He was one of Didario's sons. Didario lived two miles further up the road. We headed up the muddy road. I could see the back end of the hill where I used to live. The road dwindled into a trail that leads to the back of my old place.

Didario was home, still denying knowing the woman. "Come on now, she was staying at your home, she is your son's girlfriend," I said in English. My compadre said it kinder in Spanish. We found out she was from Orange Walk, but he did not know her name. "She is an up and down girl," explained Didario, using a euphemism to cover his disdain.

Another son returned with the truck. He would not look me in the eye and went inside the house. It would have taken two persons to load a truck to move all my things. Didario told us that his son worked the cane fields around Orange Walk and would be back next week. I asked him to have his son contact me.

Back at the police station, I passed along my findings, including my observation that none of my belongings were at Didario's. They wanted to know how much the bag was worth and when had I reported the theft. They told me that they would take care of it. They will extradite the man from San Pedro.

GOLD BAG UPDATE

January 11, 1995
Yesterday the police came by with a solar panel, some blankets and other items they had taken from Didario's place. They were not mine. I told the police that none of the items there were mine, but they had gone there and confiscated things anyway. I apologized to Didario and his family.

Before I could leave, the American farmer, who Didario worked for, came to the office wanting to know what was going on. I filled him in and then he told me how the police had come charging out there and ransacked Didario's home, upsetting everyone. Didario and his wife had moved from Guatemala to be away from such tactics.

The American advised me that Didario was a hardworking, honest person and he trusted him and his wife completely. I assured him that I did not suspect Didario. What I had asked was to speak to his son so we could find the woman with the golden bag. I did point out that if Didario had been straight with me about the woman going to Belmopan, we could have caught up with her on the Shaw bus rather than chasing the other bus to the Guatemala border.

ON THE TRAIL OF THE GOLDEN BAG

January 21, 1995

Acting on a tip of where I could find the boyfriend, of the woman with the golden bag, I borrowed a car and ventured up the Northern Highway. At 3:00 P.M. I reached Sandhill where I was supposed to go to a service station for directions to a French-Canadian. At the Shell station I was directed down a narrow road which curved left, then curved right, then curved left, then curved left again, then curved right for three miles, to a place with a fence and lots of coconuts.

The "guide" made me repeat the instructions, with all fourteen curves, back to him. It was not like the road was going to lead anywhere else!

It was the road to Altun Ha. I figured if I do not find the French Canadian I could travel the next eight miles and see the most visited Maya site in Belize. There were lots of places with fences and coconuts. After three wrong stops, I finally caught the fellow at home. He was not from Quebec but from Vernon, BC. He ran a salvage operation and sold it a year ago to retire to Belize. I was invited in for a rum and coke. The boyfriend was there.

After I explained why I was there the boyfriend "took" me to his girlfriend's place. As we continued along the road to Altun Ha, he told me that he had only known her a few months. She was 22, from Honduras; and gave me her name. He was quick to add that he was "hearing things" and would like to clear up this business of where she got the golden bag. As we became closer to Altun Ha, I wondered if I would get to see the place.

He guided me right to the site. From the parking lot, he led me through the site. It consisted of a number of large squat-type pyramids. Altun Ha is known for the largest jade head carving found in Mesoamerica. Its likeness is on the Belizean one-dollar bills. At the back of Altun Ha, there was a trail leading into the bush. The further we followed it the narrower and muddier it became. After the first half mile, I decided if my waterbed was there, it could stay there.

As we threaded our way around puddles and ponds, I noted numerous "Y" junctions. After another mile I made up my mind if someone had packed my furniture in here I did not want it back bad enough to pack it out. The boyfriend claimed that this was where he would walk her home, but he did not seem to know the way very well. I wondered how he met up with her to go on a "date" or invite her to Cayo.

We came to some shacks. This was a community of Salvadoran and Honduran refugees. These people were poor. If they had any of the things, I was going to let them keep them. No one knew the girl.

CUMBIA, SUGAR CANE AND ANOTHER WORRY

The boyfriend said he knew of a bar in Orange Walk where she might be, called *Rey Todos* (King of all). We continued along the old highway through some home sites, but the thirty-mile strip was surprisingly void of settlement. When we had to start dodging truck and trailer loads of sugar cane heading to the factory I knew we were close to Orange Walk. The boyfriend directed me to the north end of town to an establishment called *Papa Gallos* (Head Rooster).

Even though it was before 7:00 pm the place was filled with couples dancing to Cumbia music (in San Ignacio the nightlife does not start until after 10:00 pm). A small band was making a big noise. My chest vibrated from an external beat. They performed good versions of the tunes I had on cassette but it was so loud I had trouble ordering beers. The boyfriend said that he was able to have a conversation with a waitress to learn that his 'girlfriend' had found another job.

We checked other bars. The next place had women dancers, with an all-male audience - no girl. Further into town we checked out Mexicali's where a group of men crowded around the bar. As a gringo, I stood out like a turnip in a carrot patch. No girlfriend. The next bar had mostly women. It did not strike me strange, at the time, but there was one woman (and only one) at each table, and no one else, except for one table with several guys. I waited outside while the boyfriend chatted with them.

One woman followed me out and struck up a conversation in Spanish. She wanted to know what I wanted. I answered "*Busco una mujere*" (I'm looking for a woman.) She looked me in the eyes and replied "I'm a woman" to which I responded, "You sure are." I shook her hand, told her it was a pleasure to meet her and got into the car saying "Adios".

Twelve bars and half a dozen beers later we were back at the south end of town, and I did not care anymore. Just one bar left, Pasadito's. I did not want any nagging doubts about not having checked one more place. The boyfriend went in alone. I could see him talking with a woman. He came out excited about the information he had received.

After determining that I could not speak Spanish, he had me to come back in and buy him a beer.

He claimed that one woman had told him that his 'girlfriend' had cut off her long straight hair, had 'painted it yellow' and went to a place in Corozal (near the Mexican border) called *Sale Si Puedes* (Leave-if-you-can). She gave her real name and advised that she lived with a drug dealer.

I do not know if any of that was true. I thought it improbable that she would dye her hair blonde. Enough "adventure". If someone had taken my furniture and sold it (at one-tenth of its worth), I was going to look at it as a redistribution of wealth. I now had another worry. How was I going to explain to my girlfriend that I spent Saturday night bar hopping - looking for a woman?

PHYSIX BEFORE DINNER

When I arrive home before dark, the kids want a swing. And that is okay even though I have just hiked in three miles from the village of *Cristo Rey* (Christ King). I can swing them around a few revolutions. But they want another swing. I made the mistake of varying it (Okie being lighter I can lift higher), so the others want a turn "just like she got". After six swings I am tired and dizzy. That's enough for me, but it is not enough for them! They badger me for a "lift". I hold out my arms; they clamp on, one per arm, and I lift them to my eye level by which time Okie cannot hang on anymore, and Aimee is too heavy to keep elevated.

Jason is the most persistent about wanting more turns. I thought I would try to explain to him why I do not have the energy. "Jason, I'm going to give you a Physics lesson" I announce. Okie and Aimee picked up on the word "give" and they wanted one too. Sure, I agreed, "I'll give you one too. Everyone come listen." Tina was setting the table.

Terry: "Gravity exerts a pull in an exponential fashion."
Jason: "What's gravity?"
Terry: "It is what keeps you on Earth."
Aimee: "Please make me a sandwich."
Jason: "What is the Earth?"
Terry: "I think you have to wait until Tina is ready." I say to Aimee. "The Earth is what you are standing on; it keeps you from flying off into space.
Aimee: "I want a pickle ... please."
Terry: "Gravity pulls you down, that is why you hit harder the further down you fall.
Jason: "What's gravity?"
Okie "Pickle please."
Aimee: "Please, I want a sandwich."
Terry: "Gravity is what makes the pickle float in the pickle jar ... never mind that. How much does Okie weigh?"
Jason: "Thirty pounds."
Terry: "How much do you weigh?"

Jason:"Between fifty and sixty."

Aimee:"Make me a sandwich... PLEASE."

Terry:"Tina says to wait. Jason, you weigh twice as much as Okie."

Jason:"I do not."

Aimee:Pleeeeeeeeeease!"

Terry:"Trust me you weigh twice as much, and because gravity pulls exponentially that means I can lift Okie four times as many times as I can lift you before I tire out. That is why Okie gets more lifts."

Jason:"But she gets ten times as many."

Aimee:"Sandwich, sandwich, sandwich, sandwich..."

Okie:"Pickle, pickle, pickle, pickle, pickle...

Tina:to the rescue "Eat your soup first."

So much for Physics for a six-year-old.

COULD IT BE A JAGUAR?

January 1995

Late one night the dogs were barking frantically. It was two nights after a big jaguar had been seen drinking from the river, one mile upstream. "It was a BIG jaguar!" said one. The local food chain works like so: house cats eat the possums and rodents; raccoons tree the house cats; dogs chase the raccoons, and jaguars maul dogs.

In the moonless night, I went out to investigate the barking. The dogs, that were unchained, circled behind me and whined pitifully. I could not see anything in the yard, and the dogs were not venturing to point out what was driving them to distraction. I wondered if it could be a jaguar. I had not seen a jaguar in the wild. Did I really want to see one at 2:00 a.m.? The fearless, courageous dogs continued to cower behind me. Just how big was big?

Postscript *February 1: The Jaguar ate a big sheep at a place upriver. That kitty must be big.*

BLACK ROCK ADVENTURE

January 29, 1995
Further, Further Up the Creek

Milli, my landlady, wanted to learn about the Black Rock Resort which was about six miles further up the Macal River. Customers at Crystal Auto asked her about the place. I had an invitation from the owner, to visit, so we made a plan to venture out. Andrew was willing to take us up the river in his flat-bottomed boat. He figured, having to wade through nine series of rapids, it would take three hours. Before the first hour was up, I realized I should have brought suntan lotion.

Back at the house, I did have some cohune oil to test in my kerosene lamp for lighting potential. I could have brought some to test for sun tanning potential. I was going to burn for not thinking of bringing it. I also ought to have brought a camera.

It was a photo journey. Along the way, we saw the thick, fallen tree trunk that Mr. John was beginning to chop out to make a makora canoe. It reminded me of the unfinished Haida canoe on the Queen Charlottes. Across from that was the point where Andrew saw the big jaguar take a drink. A little further up we could look under the base of a giant Guanacaste tree whose maze of roots were revealed where the river bank had washed away. It looked like something out of Lord of the Rings.

Picturesque

Just past where we usually stopped to get oranges from Negroman, was a collage of vines like some artistic wall hanging draped from the treetop sixty feet above to the water surface below. Sunlight glinted through the leaves, and vine inter-wrappings like a cacophony of illumination played in melody. From the shaded coverings of the hills and trees along the banks we floated into a wide valley opening broken only by occasional, but lengthy, sets of rapids.

We could see how the road, that paralleled the river, had dwindled into a trail - passable only with 4WD in the rainy season. Andrew told us who owned which piece of land, how much it was and who had been trying to buy it. The rugged peaks of "true" mountains lured us onward. We worked our way through the next set of rapids which involved me pulling and Andrew pushing (Milli sat in the middle like some Pharaoh). The branches of a large fig tree overhung the river framing one of the limestone formations in a picturesque composition.

Further up, a fig tree disclosed several iguanas sunbathing on its uppermost branches. Yes, there were lots of birds: black ones, red ones, small ones, and big ones.

River Resorts

By land, it was a twenty-minute hike to Black Rock Resort or a boat ride away to Ek Tun Resort across the river. Ek Tun (black rock in Mayan) was a resort, with only two cabins.

I had worked with one of the owners on some local conservation issues. We dropped in to say hello and look around. The site was well groomed, and the facilities were spread out to give a spacious feeling. Their generator was hidden in a small shelter; their lawnmower was likewise concealed. Only the cook was there to greet us. The owners were conducting a tour. We enjoyed the magnificent view of the sheer, white-faced mountain cliffs and were lulled by the sounds of a set of rapids we would be traversing.

Although deep, we were able to negotiate past the falls without too much trouble by walking along the edge. (Queen Cleopatra helped by getting out.) Along here the rock was black (hence the resort names). Locals were collecting pieces of black slate to be used for carving to sell to tourists. We collected a large flat piece for doing the laundry. We reached as far as we could; deciding the next set of rapids was beyond our energy and skill level. Onshore we ate the lunch and then rested in the shade. After a quick swim, we continued on. In less than five minutes we were at Black Rock Resort.

Caesar had described his place as offering better scenery than Ek Tun's, but his accommodation was less attractive. There was a small waterfall centered in the river below, and mountains rose dramatically on all sides. Except for a propane fridge, most everything else was solar powered.

From Black Rock one can take day trips hiking or horseback riding, to Vaca Falls or to the newly rediscovered Flour Camp Cave. They billed this as being larger than Rio Frio Cave with more beautiful Maya artifacts than Che Chem Cave. From the mountain that we had used as a marker to guide us, one could look down on Xunantunich. We were impressed with the different quarters.

Prices ranged to suit backpackers or luxury seekers. What they got was No Television, No Radio, No Magazines, No Newspapers, No Traffic, and No Noise except for the call of birds and howls of monkeys. It was certainly a place to get away from it all.

JANUARY SUNBURN

On the way back, we stopped at Negroman to pick oranges. We could make out the Maya ruins of Tipu rising among the orange groves silhouetted against the setting sun. We did not see Cagney, the vegetarian cat. Back at the farm, I found that the cohune oil would not burn in my lamp. Not only did I have sunburn from not having any oil with me; I had no light burning from having tainted oil.

GARMO AND JERRALDO

February 4, 1995

If I reach Cristo Rey before dark (three and a half miles after San Ignacio, three and a half miles before home), I can usually help Garmo and Jerraldo carry water to their house. They do their water run every morning and every evening. They live on the south end of the community about four blocks up the hill from the water pump. By their size, I figured they were three and four years old. They are shorter than Okie, who at three years old is 39 inches tall.

Always full of smiles they would give me a big wave if I should happen to get a lift in the back of a truck. But I have to walk many times and was able to enjoy their company as I carried their water up the hill. Carrying a half-gallon container in one hand and 32 oz. in the other, they have to go in stages, stopping about eight times before arriving home. Garmo and Jerraldo really like it when I carry the water for them. Bubbling with enthusiasm, they carry on a conversation in Spanish.

One day, when I had the CCDC truck, I gave their mother a ride. I did not know she was their mother. When she had me stop in front of the thatched building, I knew so well, Jerraldo and Garmo came running out of that place faster than I would have thought their little legs could carry them. They were so excited about Mom getting a ride in a truck that they did not notice I was the driver.

The next time I had the truck they had just started up the hill with their water. I stopped to give them a lift. Putting the truck in reverse, I turned off the engine and put the parking brake on. Leaving the driver's door open I moved quickly to load them into the truck. They wanted me to take the water first. When I went to pick them up, they had run around to the back of the truck to climb on the bumper. The truck started slipping.

I ran around to the back of the truck and tried to hold it from moving. They were not big enough to climb in on their own. There were not too many options. The boys would not understand my English if I cried out for them to get out of the way and my Spanish did not include that essential vocabulary.

Going back and stepping on the brake may have been too late for one of them. They would be better off in the truck and I would rather it was me than them under the wheels. As the truck kept slipping in reverse gear, I put one then the other in the back.

The open door met me halfway as the truck accelerated its roll. My timing was accurate, and I caught the brake then started the truck. I delivered the boys and the water to their home. They were delighted with the ride, and I was very relieved that no harm came from my carelessness.

Last night, once again, I had the pleasure of seeing their smiling faces and carried their water. Thankfully that is the only burden.

*Postscript 2001*Jarmo and Jerraldo still haul water. One is five years old; the other is six. They have graduated to carrying quarts in each hand and are joined by their sister Claudia who is two years old. She also carries one quart of water.

Postscripts 2019:The water pump stands silent like a sentinel of time gone past. Due to the dam, the river waters is no longer drinkable. Locals no longer bathe in the river nor use it for doing their laundry. The water causes uncomfortable itching. Jarmo and Jerraldo are grown up now with children of their own. The family had moved to San Antonio.

TITBIT

When they conducted the environmental survey before building the dam on the Macal River, they learned that one of the species in the river was a type of piranha. They do not do a lot of damage, but they've got the teeth.

As the engineer, for the dam, sai: "They know what to do, they are just not big enough to do damage." Chicleros used to jump in the river after a long day's work to let these sardine-size fish bite off ticks. I discovered their nippiness while having my daily swim. One chomped on my left nipple. Ouch!

Postscript 2019: *Most of the fish in the river are poisoned with mercury due to the dam upriver affecting the vegetation to release that heavy metal into the water.*

COATAMUNDIS

While heading up the trail and carrying out some luggage for my return to Canada, I was "greeted" by two large coatimundi's racing after one another right towards me. Coatimundi's are unusual in that they look something like a monkey, something like a raccoon, something like an anteater, and something like a mongoose. Before this encounter, the most I had seen of a bush dog was its identifying tail as it disappeared into the woods. Their very bushy tails are almost as long as their bodies. These two had white-haired heads to contrast against their black bodies.

When the first one reached me, it looked up in disbelief at something so huge in its playground. It turned to warn the other but as number two caught up; it pounced on number one, and a battle ensued. In the scuffling and chattering, I was forgotten. Again they ran towards me. Just before dashing between my legs the first one turned off and dived into the bush. The second one, bewildered at the rapid disappearance of the first, and the apparently sudden appearance of me, stopped in shock at my presence. It looked me up and down in bewilderment before deciding that it too had better make for the bush. They reappeared further up the path.

GIRLS ARE SMARTER THAN BOYS

Twice a week I make two batches of popcorn. Everybody likes popcorn: the kids, the dogs, the cats, the pig, the chickens, the ducks, and even the parrot. Giving it out made for quite a following. Having purchased fifty pounds of kernels, I could afford to be generous.

At school, Jason is learning to add and subtract. Aimee is supposed to be in school also, but there was not enough room in the class, so she attends preschool with Okie. To help her get caught up, she is encouraged to work with Jason on his homework assignments. To reinforce their learning, I get them to make their own change when playing monopoly. Even Okie, at age three, could count out the number on the dice and then move her marker the correct number of spaces.

The girls sometimes complained that it is too hard for them to have to do the same things as Jason. "Because Girls Are Smarter Than Boys," I told them "You should be able to keep up with Jason." To prove my point I took some popcorn outside where dogs and cats gathered.

The cats did not get much popcorn because they had to sniff and inspect each piece. By the time they did that, and before they get it to their mouth, the dogs have swiped it. Dogs inhaled on impact. Two of the dogs were half a year old, just past being puppies. Princess catches the popcorn when thrown to her.

Bear lets it bounce off his nose. Usually, a chicken gets it before he could nose around on the ground to find it. He spent most of his time looking for disappearing popcorn.

I pointed out "See Bear is stupid, he hardly gets any popcorn." And then I throw to Princess "See! She catches it most of the time. Princess is a girl. Girls are smart. That is why she catches the popcorn and Bear doesn't." I point out. "Now you might think my database is limited," I continued as we strolled to the dogs posted at corners of the farm.

"Here is Muffet, she is a girl. See how she catches popcorn in her mouth? She is smart." Then to bolster my argument, I toss popcorn to Princess and Bear again. "See. Bear cannot catch the popcorn but Princess and Muffet can."

For more empirical evidence we went to the other side of the house where Beauty has her run.

She came bounding. I repeated the demonstration. "See how Beauty catches the popcorn?" Bear, Princess, the chickens, and the ducks have dutifully tagged along. Again I show how Princess catches, and Bear does not. The "guys" don't get it.

I ask "Okay, who is smarter?" Okie and Aimee answered in unison "Girls are!" Then I asked; "Now you won't have any problem with your homework will you?" They reply "Noooo!"

"That's good," I told them. As we walked back, just to show there are exceptions, I threw a piece of popcorn high in the air - and caught it in my mouth.

Postscript *2019: Aimee graduated with honors in Massage Therapy. Okie became a nurse and is working in Dallas, Texas. Jason is also working in Dallas.*

DAM AGAIN

February 19, 1995

In the story titled "DAMn" I wrote on the environmentalists' perspective of the Macal River Dam. Ron Shaw, Chief Engineer for the Dam Project had invited CCDC staff to take a tour. We were at the designated place that Sunday morning but no Ron. That was not like him. We doubled back to Clarissa Falls where he lived.

Chena advised that he had gone to town but would be back soon. She added that it was not like him to forget, but perhaps he had. We had breakfast by the falls, and sure enough, he returned from his weekly shopping. I went over to greet him, and he gave me a big smile and welcome. "What's up?" he asked. He thought I had brought more books to exchange.

"We came for the dam tour" I informed him. His brow creased into a frown and then he announced: "I forgot."

"No worries" I replied. We left immediately for the dam place. Along the roadway as we passed the turnoff to Che Chem Cave and could discern the distant cliffs of Black Rock. The further south we travelled the more dramatic the landscape. We stopped at an overlook to enjoy the spectacle of a distant waterfall.

Ron explained how they had built a concrete wall at this turn-off because truck drivers faced with a sheer drop tended to panic rather than make the turn. Behind that waterfall was a three-mile trail that led to Augustine. The hills lay in a raw beauty of rainforest carpeting.

We continued to the end of the dam site. By the time we arrived Ron was already speaking to one of the dam crews about fixing a dam leak in one of the dam settling tanks. He explained that part of his dam job was to ensure that dam environmental concerns were addressed.

The stone crushing mill was at this location. It worked out to be cheaper to make their own dam sand from the granite they took out of the dam tunnel than to haul sand from Bullet Tree Falls (the closest suitable sand). It also cut down the number of trucks trips from 100 to 50. Large yellow cylinders served as giant cement mixers for the dam construction. I wondered if they would stay for other projects.

Ron elaborated on the financial benefits to Belize by having their own power source rather than paying for imported diesel fuel to run polluting generating stations. We moved 400 yards further downstream.

It was a relatively small dam as far as dam projects went. He pointed out how the sides of the river had been cleared to about flood level, which was also the level of the dam. The dam lake would be about a mile long. It would have a reserve for up to seventeen hours of water. That means if there is an extended drought the turbines would not run.

Out of several scenarios examined this dam undertaking was the deemed the one with the minimum environmental impact by a study (financed by the Canadian International Development Agency). The main component, costing about 40% of the whole project was the two-mile-long tunnel. Granite blasted out of the tunnel was the source for making the sand. Environmentalists are concerned that when that inevitable drought occurs people will be clamouring to build another dam with a much larger lake and much larger environmental impact.

WOULDN'T STOP A FLOOD

They had estimated that the dam could hold back a flood for eleven seconds. It was not a major dam. The main feature was the two-mile tunnel by-passing a six-mile oxbow in the river. It allowed for a drop of 400 feet, which created the thrust to turn turbines to generate electricity. The next stop was the dam itself.

Lovely old era rock was exposed from being hidden would now be permanently under water. On the other side is the beginning of a stretch that will be without its water source. The tunnel began on the left side of the dam. There was a sharp turn before making the long run to the powerhouse.

TUNNEL IN

About a half-mile beyond the dam is a way into the tunnel itself. Drilling teams alternated working opposite ends of the tunnel, thus this intermediary access point. They worked two ends of the tunnel at the same time. The drilling team drilled six-foot holes in a circular pattern. Then the dynamite team packed the explosives and ignited the charge. While they were doing that, the drilling team went to the other end. After the rock tumbled down the third crew cleared the rubble, and they repeated the process twice a day.

Another mile further down, the same approach was taken - coming in from the side and then drilling out to meet the first tunnel on one side and to the generator on the other. Thus the tunnel was built in four segments with only two of them having to meet up.

Although the slope is gradual, it was a surveying feat to both measure over the variable altitude of the rugged hills to make the ends align within six inches. The difference was not noticeable in the roughed out exploded surfaces.

It was two vehicle widths wide and about thirty feet high. Anything bigger would have to be shorn up inside with concrete. It did not seem to me that the river carried enough water to keep the tunnel filled (in normal flows). Despite it being lit inside we could not see either end. We were unable to drive through the tunnel as they were still working at the far end.

DYNAMITE SHACK TO DISAPPEAR

Along the road back to the powerhouse Ron pointed out the facility that was used for storing the dynamite. Then he pointed out the hidden Belize Defense Force observation post. It was designed to be undetected in the rainforest.

Those facilities, along with the construction sites, camps and equipment shops, would all be cleared out. In three years, signs of use would no longer be visible due to rapid forest regeneration. The road would stay.

POWER

The powerhouse consisted of an office on the top level with two large generators. One was mounted on railway tracks so that it could be moved. The other was a spare; they could be switched over in four hours. Giant turbines were being installed down below. Although twelve stories high the complex was dwarfed by the sheer cliffs of the rugged mountainside.

It was yet to be seen what the overall environmental impacts and financial benefits would be. It will be left to those in power to determine the future. Whatever the direction, it is claimed that the dam attraction with waterfall viewing, hidden Maya sites, and fabulous caves will offer tours to rival the Mountain Pine Ridge route. For those hiking, or on mountain bikes, they will be able to do a circle route through both. This remains to be seen.

Postscript 1995: *It did not stop a flood. In August of 1995, the Macal River overflowed its banks reaching San Ignacio's main street.*

Postscript 2001 *The government plans to go ahead with another bigger stage of dam development. Johnathan Lohr is working with environmental groups to make Belizeans more aware of the issues.*

EPILOGUE

There are now three dams on the Macal River. The first was the Mollejon Dam mentioned in the previous story. Then came the Chalillo Dam, known as the Macal River Upstream Storage Facility. It created a large reservoir of water for use by the Mollejon Dam that was never capable of producing the electricity it promised. This second dam caused the most environmental harm. The third, Vaca Dam, was built between the previous two.

The dams were built to provide electricity. In doing so, they had immense negative consequences. When I lived in San Ignacio, I recall the promise of reduced electricity pricing and no longer having it generated by diesel power. The diesel engine operation ceased, but the price has escalated since the dam was constructed. In the early '90s electricity cost 6 cents per KwH. It is now 34 for the first 50 KWhours, then goes up to 40 cents/KWh. It is among the most expensive electricity in the Americas. We were told the price would go down with the building of the dam. It seems proponents could say anything they wanted and would not be held accountable.

Hydro accounts for 28% of electrical generation in Belize. Some are extraneous to the dams such as the run-of-the-river project providing electricity at Blancaneaux Lodge. Approximately 20% is renewable energy, and the rest is fossil fuel generated (much of that purchased from Mexico). For all the damage and repercussions of the dams, they only meet one-quarter of Belize's needs.

The bike trail circuit envisioned by Ron Shaw never materialized. There may have been a "hidden" Maya site that has been accessed, but there have been other sites submerged by the flooding, under objections of the Archaeology Department and archaeologists working in the country. There is recreational access to the lake made by the dam, but it is mostly for visitors staying at the high-end resorts.

Canoeing options on the river are close to non-existent due to erosion caused by the daily release of water that has been held back which means repetitious decomposition of the vegetation. The loss of canoeing and other tourism-related activities (such as guided expeditions) has been a tragedy for the communities. Employment opportunities have been lost to the detriment of those who had provided for their families. They are now impoverished.

Community life in the villages further denigrated. Women and children used to congregate at the river to do laundry collectively; to talk, to laugh and have their children play together. That social aspect of village life disappeared because the quality of river water diminished. Due to itching, that lasts for months, people no longer do their laundry there, nor do they swim in the river.

Where I used to stay, I could no longer swim nor simply wade across the river to visit Chaa Creek. My canoeing trips are no longer feasible. The water is not drinkable; it is so bad that even livestock are kept away from the river. The turbidity levels of the water prevent the water from being filtered for drinking. Drinking water is now bottled water purchased in town at expense and inconvenience to transport instead of pumping from the river. There is no more fishing in the river. Mercury levels have risen so high that residents cannot eat the fish. The increased heavy metal levels are attributed to biological decay from the dam construction.

Sediment is held back by the dams. When there are floods, they are more devastating than before due to the lack of sediment that would normally slow down the movement of water. Thus the erosion is far greater. The Chalillo Dam was built on sandstone instead of granite, which makes it more susceptible to cracking, particularly since it is situated along a minor fault line. There is no workable emergency evacuation plan. There is no viable warning system in place if the dam breaks. Vegetation has also been destroyed upstream from the dam. Wildlife has been affected. Scarlet Macaws have experienced severe habitat loss. Their numbers have diminished. A small and struggling population has been trying to keep the Chalillo dam area as a home.

The macaw's natural breeding and foraging habitat have been disfigured. A scarlet macaw will keep returning to its nesting site, even though it is sub-standard for their reproduction attempts. These rare birds have become geographically isolated and are now faced with poaching due to easier access to the area. Warnings of the parrots' demise were given during the Chalillo dam hearings

Sadly Belizeans may only know a scarlet macaw from a zoo visit. For a fascinating account of the battle to save these birds read:

The Last Flight of the Scarlet Macaw: One Woman's Fight to Save the World's Most Beautiful Bird by Bruce Barcott.

It is a remarkable story of not only the biography of the founder of The Belize Zoo but provides insights into how corporations get their way and how the legal system has been thwarted from bringing justice.

APPENDICES

Mileages: Belize City to Caracol and Tikal

A caution about drivers in Belize: They pass on the right; they pass on the left, they pass while you are stopped for a pedestrian; they pass when there is a double yellow line. They pass in school zones; they pass on curves and they pass at speed bumps. They pass when there is oncoming traffic; they pass when you are speeding and they pass when you are passing.

Appendix A   **Belize City to Belmopan** (one hour)
(distances given in miles)

Mileage	Highlight	Travel Time
0.0	Start at Crystal Auto Rentals	0 minutes
1.9	Traffic Circle to Belmopan	5 min
2.8	"End of City" Traffic Circle	7 min
5.4	Old Belize Attraction	11 min
6.4	Burden Creek Bridge	13
8.1	Community of Eight Mile	16
15.3	Traffic Circle to Burrell Boom	20
15.6	Hector Creek Bridge	21
16.2	Hattieville Police Station	22
19.3	Fu Wi Flavaz Restaurant	27
29.2	Belize Zoo	40
29.7	Colonel English Bridge	41
29.9	Coastal Highway turnoff	41
31.5	Cheers Restaurant, then Amigos	42
37.3	Turn off to Cave Tubing	49
38.1	Community of St. Mathew	50
39.4	Beaver Dam Bridge	51
42.4	Community of Cotton Tree	56
46.3	Kiki Witz Wellness Resort	59
48.2.1	Belmopan Traffic Circle	1 hour

Appendix B   Belmopan to Cayo* (45 minutes)

*San Ignacio and Santa Elena together are known as Cayo

Mileage	Highlight	Travel Time
48.2	Guanacaste Park	1 hr 1 min
48.3	Roaring Creek Bridge	1:01
48.4	Community of Roaring Creek	1:01
49.5	24 Hour Gas Station & Hotel	1:05
50.4	Community of Camalote	1:08
53.2	Community of Teakettle	1:14
53.3	Actun Tunichil Muknal Maya site	1:14
53.8	Detour (because of construction)	1:15
54.7	End of Detour	1:18
54.8	Trey Barn & Grill sign	1:18
55.5	Valley of Peace Farms	1:19
56.0	Ontario Village	1:20
56.7	Jingle on the River Resort	1:21
57.6	Community of Blackman Eddy	1:23
58.2	Turn off to Spanish Lookout	1:24
58.3	Yellow pyramid Temple	1:26
59.4	Community of Unitedville	1:27
60.2	Timber Cube (crafts for sale)	1:28
61.1	Orange Gift Shop formerly Caesar's	1:30
62.0	Redmonds Garden Nursery	1:32
63.2	Barton Creek Caves Turnoff	1:34
64.6	Central (Government) Farms	1:36
64.7	Alternate road to Spanish Lookout	1:36
66.8	Community of Esperanza	1:40
67.3	Town of Santa Elena	1:41
68.5	Traffic Circle to San Ignacio	1:42
69.3	Mountain Pine ridge turn off	1:44
	(For access to Caracol)	

Appendix C   **Cayo to Caracol** (two hours)
                    (distances given in miles)

Mileage	Highlight	Travel Time
0.0	Turn off Western Highway	0 minutes
0.7	Maya Mountain Lodge	2 minutes
1.2	Cedar Bluffs	3 minutes
1.9	Devil's Hill	5 min
2.6	Village of Cristo Rey	6 min
4.6	Table Rock Camp & Cabanas	7 min
5.1	Entrance to Mango Walk	8 min
6.0	Mystic River Resort turnoff	9 min
6.6	Magna's Zactunich Art Gallery	11 min
8.3	Garcia Sisters' Museum	14
8.7	Speed bump (1of 8) San Antonio	18
9.4	Bakery on right, a church in front	19
10.6	Sign *"Next Tim Horton's 4,888 km"*	22
11.2	Pacbintun Maya site	23
12.5	Mariposa Jungle Lodge	25
12.6	Junction to Caracol turn right	26
12.7	Moonracer Farm (closed in 2019)	27
14.1	Gate to Mountain Pine Ridge	30
17.6	Road to Hidden Valley (1,000') Falls – 10 miles distant to left	37
17.9	Pine Ridge Lodge	38
18.1	Turnoff for Blancaneaux Lodge	43
18.7	Privassion Creek Bridge	47
21.0	Oakburn Creek Bridge	53
22.5	Pinol Sands rest area	63
22.7	Pinol Creek Bridge	64
25.2	Rio-on Creek	67
25.3	Rio-On Pools and views	70
27.6	Domingo Riz Cave turn off	73
28.0	Village of Douglas DaSilva (Turn right for Rio Frio Cave) (Turn left for Caracol)	74
~38	Cross the Macal R. at Guacamallo	1 hr 20 min
~51	Ancient Maya Site of Caracol	2 hours

Appendix D　　**Cayo to Guatemala Border** (45 minutes)
　　　　　　　　　(distances given in miles)

Mileage	Highlight	Travel Time
0.0	San Ignacio Police Station	0 minutes
0.2	San Ignacio Hotel	1 minute
0.7	Victor Galvez Stadium	2 minutes
0.8	Traffic Circle at Double Street	2 min
1.4	Kon Tiki Gas Station	4 min
1.7	Cassia Hills Resort	5 min
4.7	Clarissa Falls Resort	8 min
5.0	Turn to Chaa Creek & Black Rock	9 min
5.8	Nabutunich Resort	10
6.4	Village of Succotz	12
7.4	Town of Benque Viejo	13
9.1	Border Crossing to Guatemala allow thirty minutes to cross	15

Appendix E   **Melchor to Tikal** (one hour, 45 minutes)
                   (distances given in miles)

Mileage	Highlight	Travel Time
0.0	Border town of Melchor de Mencos	0 minutes
3.4	Sign for a Hotel	8 minutes
7.1	Turn off to Sabintos Village	10 min
12.2	Village marker (at Km 537)	19 min
19.3	Yaxha-Naranjo Park turn off	29 min
20.6	Sign for Holtan	39 min
39.2	Right turn for Tikal (Flores to the left)	1 hour
40.5	Village of El Remate	1:02
42.5	Village of El Copuliner	1:05
45.3	Village of El Cabba	1:08
47.5	Village of El Porvenir	1:13
51.1	Gate to Tikal Park (Pay fees)	1:22
61.8	Reach Tikal complex site	1:44

To reach Uaxactun (a combined village & protected Maya site)

62.0	Turn left at traffic circle	1:45
71.2	Sign that says "10 Km to Uaxactun"	2:05
72.0	"End for Norte" sign	2:10
74.2	Sign that says "5 km to Uaxactun"	2:15
77.7	Uaxactun	2:30

Appendix F
CHRONOLOGY OF **BELIZEAN SITES**

Maya City	Pre-Classic [	Classic Period	] Post-Classic
Altun Ha	1000 BC _____		___] 1000
Arenal (Las Ruinas)		*data not found*	
Buena Vista		*data not found*	
Cahal Pech	1200 BC_____		___] 800
Caracol	650BC[_____		___] 950
Cerros	4,000BC[_____	]AD 400	
El Pilar	800 BC[_____		___] 1000
Lamanai		AD700[_____	]AD 1000+
Lubaatun		AD700[_____	]AD 890
Nim Li Punit		[_____	]AD 790
Pacbitun	900 BC[_____		___]AD 900
Uxbenka		[_____]	
Xunantunich		AD600[_____	]AD 890

Appendix G
CHRONOLOGY OF **GUATEMALAN SITES**

Maya City	Pre-Classic [	Classic Period	] Post-Classic
Bonampak		AD 580[_____] 800	
Dos Pilas		AD 629[____] 761	
El Mirador	700BC[_____]AD 200		
El Peru	500 BC[_____] 800		
El Zotz		_____]744	
Kaminaljuyu	1500BC_____ 1200]		
Naranjo	500 BC [_____] 820		
Quirigua		AD 200 [_____] 900	
Seibal	400 BC [_____] 890		
Tikal		AD 200[_____] 900	
Uaxactun		AD 378 [_____] 839	
Yaxchilan		AD 359 [_____] 808	
Yaxha	1000 BC _____] 900		

Appendix H
CHRONOLOGY OF **MEXICAN SITES**

Maya City	Pre-Classic [	Classic Period	] Post-Classic
Calakmul		AD 300[_____] 700	
Chichen Itza			AD 600[_____] 1200
Coba	50BC[_____]1500		
Mayapan			AD 600[____] 1170
Palenque	250 BC[_____]1000		
Uxmal		AD 500[_____]1000	

NOTE FROM THE AUTHOR

Word-of-mouth is crucial for any author to succeed. If you enjoyed the book, please leave a review online—anywhere you are able. Even if it's just a sentence or two. It would make all the difference and would be very much appreciated.

Thanks!
Terry

ABOUT THE AUTHOR

After graduating from the University of Calgary, Terry Vulcano worked five years for CP Rail in Montreal, before going to Africa as a volunteer for three years. While there he became interested in pre-historic movement of people and returned for a degree in Archaeology. Then came an opportunity to go to Belize to help cooperatives with their economic development projects. About once a week he would write about incidents that happened.

Thank you so much for reading one of our **Humor Memoirs**.

If you enjoyed our book, please check out our recommended title for your next great read!

Nazis & Nudists by David Haldane

"Haldane's storytelling is rapid, fact-packed, devoid of filler (and) heavy on action." -*Long Beach Press Telegram*

"The story is unflinching, wildly improbable and pretty scary in spots." –Ken Borgers, *KSDS 88.3 San Diego*

CPSIA information can be obtained
at www.ICGtesting.com
Printed in the USA
LVHW010405201119
637945LV00002B/2/P